FURTHER PRAISE FOR
TIME TO THINK SMALL

"Todd Myers is the voice we need in the conversation about mitigating climate change risks. His research and insights prove to be valuable time and again, steering us towards free market solutions that will get results."

Cathy McMorris Rodgers

"Myers, known for his provocative opinions, takes a turn toward absolute and undeniable common sense in his demonstration of the opportunities now readily available for addressing seemingly intractable environmental issues with devices and technologies that are available to most of us, right now."

Sheida Sahandy, former director of the Washington
State Puget Sound Partnership

"*Time to Think Small* makes a compelling case that grassroots efforts are powerful tools for dealing with environmental challenges. By providing and sharing information, while affording control of personal choices and options, great progress is possible—using smartphones, personalized AI, distributed sensors, and wireless communication—in dealing with issues from global warming and air pollution to water quality and protecting natural habitats."

Cliff Mass, atmospheric scientist,
University of Washington

"Myers's first book on 'eco-fads' outlined the lazy, troubling, and increasingly prevalent tendency of public figures flocking to 'green' and trendy environmentalism, rather than focusing on solutions truly rooted in science. Now, he invites readers to explore how advanced technologies are empowering individuals to achieve real results for the environment. Myers continues to demonstrate why he is one of the sharpest and most effective communicators on environmental solutions, not just in the Pacific Northwest—but in the country."

Dan Newhouse

TIME TO THINK
SMALL

How Nimble Environmental Technologies Can Solve the Planet's Biggest Problems

TODD MYERS

imagine!

An Imagine Book
Published by Charlesbridge
9 Galen Street
Watertown, MA 02472
(617) 926-0329
www.imaginebooks.net

Library of Congress Cataloging-in-Publication Data
Names: Myers, Todd, author.
Title: Time to think small: how nimble environmental technologies can
    solve the planet's biggest problems / by Todd Myers.
Description: Watertown, MA: Charlesbridge Publishing, [2022] |
    Includes bibliographical references. | Summary: "A guide to how small
    technologies can solve environmental problems in ways unimaginable
    just a decade ago."—Provided by publisher.
Identifiers: LCCN 2021049612 (print) | LCCN 2021049613 (ebook) |
    ISBN 9781623545543 (hardcover) | ISBN 9781632892485 (ebook)
Subjects: LCSH: Environmental engineering—Technological innova-
    tions. | Green technology. | Environmental protection. | Environmen-
    talism.
Classification: LCC TD153 .M94 2022 (print) | LCC TD153 (ebook) |
    DDC 628—dc23/eng/20211209
LC record available at https://lccn.loc.gov/2021049612
LC ebook record available at https://lccn.loc.gov/2021049613

Display type set in Core Circus by S-Core, Aboriginal Sans by Chris
    Harvey, and Avenir Next by Linotype
Text type set in Baskerville by Monotype
"Earth" by Kevin M. Gill is marked with CC BY 2.0
Printed by Bang Printing in Brainerd, Minnesota
Production supervision by Jennifer Most Delaney
Jacket by Nicole Turner
Interior design by Mira Kennedy

Printed in the United States of America
(hc) 10 9 8 7 6 5 4 3 2 1

For Bob Kahn, who shared my love of innovation.
May his memory be a blessing.

Maria, my muse and advocate.

My colleagues at the Washington Policy Center,
who challenge me and laugh at my jokes.

CONTENTS

FOREWORD

Tromping through the steep, poison oak–infested mountains of my hometown, I held a clunky metal antenna above my head and paused to listen. A faint *beep, beep, beep,* and my heart leapt—after hours of deafening silence from the device that I could have sworn was older than I was, we were getting close to a young male puma on the move. It was the summer before my junior year of college, and I was interning with an ecology lab at the University of California, Santa Cruz. I spent most days like this, attempting to track down individual pumas to get close enough to download data from their radio collars, with painfully mixed success thanks to the thickly forested terrain. As an avid hiker and aspiring wildlife conservationist, I didn't begrudge the work, but with an iPhone in my back pocket that could tell me the exact location of the nearest Pokémon or the weather in Nairobi, I repeatedly found myself thinking *there must be a better way.*

As it turns out, I was not alone in that feeling. Two years later I joined the staff of **WILD**LABS, a nonprofit collaborative and vibrant global online community of like-minded engineers, makers, and conservationists working to build and implement better environmental technologies. Now I lead the **WILD**LABS research program and get to spend my days working to help conservationists access the resources and support they need to leverage these rapidly advancing tools.

This was the story I shared with Todd Myers when he asked how I got involved in conservation technology during an interview for this book. I also shared that my frustrations in the field were far from unique. Conservation scientists all over the world have similar experiences of this juxtaposition—watching as technologies make every aspect of our lives more convenient, efficient, and economical, yet having only outdated and expensive tools with which to protect and conserve the ecosystems that quite literally sustain us.

In the pages of *Time to Think Small,* Myers provides a timely and insightful investigation into the power of small technologies to revolutionize environmental stewardship, raising reasons for hope amidst the overwhelm experienced by so many in the face of today's environmental challenges. One of the central messages throughout the following chapters is that much of conservation technology's power lies in decentralization. Myers demonstrates that personal technologies present the opportunity to move beyond our historical reliance on top-down political solutions and into an era of harnessing the collective power of many small efforts. As a beekeeper, he likens these efforts to individuals in a hive, all contributing to something larger than themselves. A big part of my job, and the reason **WILD**LABS was created, is to connect and support these individual solutions, bringing them into a collaborative ecosystem much like a hive, where together they can produce remarkable results. While empowering individual efforts does foster innovation and new ways of thinking, without proper support systems it can also lead to duplication of efforts and competition for limited funding in an already under-resourced arena. Like any hive, it needs communication, collaboration, and resources to thrive.

Similarly, the solutions put forward are of little use if they are not accessible to the communities that need them most. As Myers details in this book, small technologies are becoming an increasingly important part of environmental stewardship, but access to these technologies remains more readily available to some groups than others. In a survey of the global conservation technology community by **WILD**LABS, 44 percent of end-users reported that technical barriers prevented them from effectively leveraging technology in their work. Along with difficulties in accessing training and resources to build those skills, financial barriers were also frequently found to reduce technology adoption and tended to disproportionately affect marginalized groups. It is therefore especially critical to engage local and Indigenous communities, the most knowledgeable stewards of the land, in conservation technology efforts to ensure equitable access to the benefits of such tools, as well as the long-term sustainability and effectiveness of solutions.

By identifying where barriers exist within current environmental technology efforts, **WILD**LABS and others can advance the field by creating pathways to collectively address them. On our platform, technology users and makers alike can engage with programs created specifically to meet their needs. From tutorials walking beginners through their first tech deployments to fellowships that provide financial support and mentoring to individuals wanting to take their work to the next level, we aim to create the space for each member to grow. Through our research program, we consistently take the pulse of progress and challenges, working to identify what will be needed next and how we can build on other sectors' advancements. As reflected in examples included throughout this book, the capacity of the ever-growing conservation technology community to forge a better future for our society and planet even with such limited resources is astounding. With access to the training and resources they need, today's independent innovators can become tomorrow's collective drivers of environmental technology solutions we have yet to imagine.

As someone immersed in a community dedicated to the very solutions that Myers outlines as beacons of hope, I must admit that even I occasionally struggle with the enormity of the challenges at hand. But readers of *Time to Think Small* can't help but come away feeling energized by the reminder that, at the very least, there are far more opportunities for positive change today than there were a mere decade ago—if only we empower ourselves and each other to seize them. That desire to find *a better way*, which took hold of me during that first summer of fieldwork, has driven me to seek out opportunities for creative problem solving beyond the boundaries of the status quo in everything I do. I can see this same drive in Myers's pursuit of alternatives to the government-based solutions he is so deeply familiar with after a career's worth of lessons learned.

As our interview wrapped up, Myers paused to ask if I had any questions for him. I had about a million but kept it to twenty minutes or so of picking his brain before getting off the phone. As I mulled over our conversation and eagerly read through the chapters he sent over in the days that followed, I realized that this was the first time an interviewer had actually made space for that kind of dialogue with me. It is that same spirit of genuine openness and curiosity that pulls me into the stories these pages hold.

Anyone curious about technology, the environment, or the future of our planet will find something of interest in the refreshingly frank and thought-provoking narrative of *Time to Think Small*. The truth is that both technological advancements and environmental challenges are here to stay as defining realities of the twenty-first century. The question then remains: can we leverage these advancements to alter the course of our changing environment before it is too late? The answer to that question will equally define this century, and Myers has no shortage of ideas about how we can approach such a monumental task. To echo the sentiment of Darlene Cavalier, who you will meet in chapter 8, may we boldly refuse to take the world as we find it as the best that it can be.

Talia Speaker
WILDLABS *Research Lead at*
World Wildlife Fund

Why Small Technology Is the Future of Environmental Stewardship

"The role of the infinitely small in Nature is infinitely great."
—Louis Pasteur

People are more interested than ever in environmental issues but increasingly frustrated that environmental problems aren't being solved. Governments propose solutions, and as each would-be solution fails, politicians turn to yet another political approach that is even more sweeping than the one that failed.

I sympathize with this thought process. In the 1970s, the United States and other developed countries faced serious environmental problems. We turned to big government agencies to address large, single sources of pollution—known as point sources—and we are fortunate they did. Our air and water are much cleaner as a result.

Today's environmental threats, however, are very different than those we faced in 1970. Who said so? Bill Ruckelshaus, the first director of the Environmental Protection Agency. Ruckelshaus, who deserves a great deal of credit for creating those early environmental successes, explained how dramatically things have changed. "Over the course of the past four decades, we have largely brought the point-source pollution problem under control," he wrote in the *Wall Street Journal* in 2010.[1] By the same token, we have made little or no progress on non-point-source pollution, like runoff from streets that carries bits of tire rubber and brake dust into the water. The opportunity to target single, large sources of pollution, as the EPA did effectively in the seventies and eighties, has largely vanished.

For two decades I have worked in government and public policy on some of the most iconic environmental issues of the day—spotted owl habitat, old-growth forests, catastrophic forest fires, dwindling salmon habitat—and I have seen many environmental initiatives come and go, with huge sums of money spent on large projects that promised to help the environment but in reality accomplished little or nothing.

We need a change. Ruckelshaus put it this way: "Yesterday's solutions worked well on yesterday's problems, but the solutions we devised back in the 1970s aren't likely to make much of a dent in the environmental problems we face today."

When we see big environmental problems, we instinctively feel the solution has to be just as big. But big solutions aren't working. Faced with the perceived enormity of environmental challenges, people feel helpless. But that need not be the case, as I will show.

It is time to think small.

We aren't accustomed to thinking in this way about environmental problems, but in just the last decade technology has dramatically changed what is possible. Smartphones and other personal technologies allow us to address environmental problems in remarkable new ways. In Bill and Melinda Gates's 2019 annual letter, Melinda Gates wrote that she was surprised at the way mobile phones have become an empowering and democratizing force, capable of reaching across the globe.[2] Increasingly, mobile phones are empowering those fighting for the environment as well.

The ability to share information rapidly and (almost) without cost is giving people the power to tackle some of our toughest environmental problems. In California, farmers are providing habitat for migratory birds thanks to data from a bird-watching app. In Nicaragua, fake eggs placed in sea turtle nests hold the promise of protecting the endangered marine animals. When poachers steal the eggs, they can be tracked by smartphones, allowing people to break up poaching networks. In Estonia, Bangladesh, and Brooklyn, smartphones allow people to generate, sell, and buy renewable energy as if they were little utilities.

Environmental activists using smartphones and personal technology are creating new solutions by overcoming barriers identified by several Nobel Prize winners in economics. There are hundreds, perhaps thousands, of examples. Connecting people to information, incentives, and feedback can effectively address environmental problems of air and water pollution, waste that reaches the ocean, habitat destruction, resource depletion, and CO_2 emissions. These new capabilities enable greater conservation and the pooling of resources, helping us make sound environmental decisions.

Sometimes it is hard to know the right thing to do for the planet. What sounds good may not necessarily be so. Rooftop solar panels, for example, are one of the most expensive and least effective ways to help the environment.[3,4] Buying local food can actually increase water pollution and waste.[5] According to research from the Danish and UK governments, plastic grocery bags may actually be better than cotton bags for the climate and for water.[6,7] You may disagree with all or some of those claims, and you may be right. It depends on your individual circumstances. If you live in Phoenix, Arizona, for example, solar panels could be a smart choice. Using your own cotton bags continuously and without exception for shopping for several years is probably better for the environment than the alternatives. Each of these choices depends on personal circumstances and behavior. The best solutions for the environment are personal.

The reason we've looked to politicians for environmental protection is not because one-size-fits-all regulations are the best option. It is because sometimes they have been the only option. I am proud of the work I did as a government employee, working for the Washington State Department of Natural Resources, and as a member of the Puget Sound Salmon Recovery Council. I know the good work government agencies can do. I also know the limitations of government and, on the other hand, have seen and researched individual, crowd-sourced solutions that have been very effective.

Political economist Elinor Ostrom was awarded the Nobel Prize in Economic Sciences in 2009, and her book *Governing the Commons: The Evolution of Institutions for Collective Action* details how local communities were successful in achieving goals for the benefit of the greater good without requiring top-down regulation.[8] One example was paying forest rangers in Japan with sake for stopping illegal logging. After all, incentives work.

Until recently, it was hard to create those small, local solutions. The cost and time needed to bring people together often outweighed the ultimate benefit. That isn't true anymore. Smartphones now empower individuals by putting direct and immediate access to information and communities in each individual's hands.

This book is about the people and businesses around the world who are using smartphones and personal technologies to create small, environmental solutions that will leave our planet a better place. Some of these efforts will fail. Some already have. But the ability to nimbly experiment and rapidly create new solutions means personal technologies will become an increasingly important part of environmental stewardship.

THE ALTERNATIVE

Quietly conservationists, consumers, and others are recognizing the opportunities offered by smartphones already in the hands of millions of people who have the knowledge and incentives to meaningfully help the environment. These solutions are appearing organically, filling gaps left by the failures of existing environmental policy. "I myself am very skeptical of technology," says Sarah Otterstrom of Paso Pacifico, an environmental NGO protecting sea turtles in Central America. She is overcoming that skepticism, noting that if we rely only on more traditional conservation approaches, "we are missing a really big opportunity at multiplying what we do."

Millions of people across the globe have embraced that opportunity. They rely on Uber, Lyft, and other ridesharing apps for transportation. People trust their neighbors' judgments on Yelp to determine where they will eat and whether to hire a contractor. We ask Alexa to answer questions. Smart thermostats such as Nest and Ecobee combine our temperature preference with artificial intelligence to keep us comfortable while reducing our heating bills. These technologies are becoming part of how we live our lives, connect to others, and discover the information we need to make good decisions. Microsoft's president, Brad Smith, highlights increasing "tech intensity" in his book about the growth of technological solutions. He believes that technologies using artificial intelligence will power "tools and devices that run almost every aspect of society and our lives."[9] That is one reason Microsoft launched "AI for Earth," to use the power of tech intensity to solve environmental problems. The ubiquity of personal technology makes it a great tool to address the myriad of environmental challenges we face. Michelle Lancaster, a member of the AI for Earth team, put it well when she told me, "Environmental problems are getting larger, and the only thing that can keep up with that is technology."

Smartphones are only part of the story. In some cases, smartphones themselves are the game changer. In other instances, the smartphones are just a tool that makes another key technology more effective. The ubiquity of personalized technology is an opportunity to create environmental solutions that were unavailable just a decade ago. We are experiencing a quiet but profound revolution in the way we think about environmental solutions. Smartphones have been a major catalyst in that mindset shift.

This is about more than convenience. Smartphones and connected technologies solve several specific problems that have hamstrung traditional political approaches to environmental issues.

The ability to harness individual actions in support of environmental protection is small technology's greatest promise. Abstract threats to sea turtles thousands of miles away are difficult to grasp and understand. People across the political spectrum care deeply about the environment, but the gap between what people know, what they can control, and their desire to help is cluttered with politics. Where people feel they can't make a difference, they demand others do so. They push politicians to enact rules that are often more symbolic than effective. Politicians target "polluters," whether that means corporations, farmers, or your neighbors who drive big trucks. Feeling small and not knowing what to do, people grasp at political and social pressure as the "force multiplier," in Sarah Otterstrom's words. The problem with relying on politicians is that you can't rely on them. What is adopted today may be undone tomorrow.

To be sure, individuals empowered with small technologies don't replace government. They do, however, engage individual efforts and resources in a way that is new and unique. Elinor Ostrom spent much of her career studying the ability of government institutions to solve resource problems. "What we have ignored," she noted, "is what citizens can do and the importance of real involvement of the people versus just having somebody in Washington make a rule."[10]

Small technologies do this in several important ways, as I will illustrate throughout this book.

At the most basic level, they provide information to make better resource decisions. Connected technologies help track water usage, find places to charge electric vehicles, connect you with car pools, and cut electricity use. You don't have to be an environmentalist to realize that saving resources also saves money. They also allow individual bits of information to be aggregated in a way that is more powerful and useful. Ecobee is using data from its smart thermostats across Canada and the United States to identify how homeowners can effectively save energy with strategies that are exact and customized. The data from thousands of individual homes, even when it is anonymized, are extremely valuable, because there is no other way to accurately collect it. Even the most expert government workers cannot craft policies that are as precise as individuals who know their circumstances and interests.

Personalized environmental efforts can also be more powerful because they align financial incentives with environmental effectiveness. By tracking my energy use, I can determine if the lightbulbs I installed are actually saving money. Smartphone apps allow me to see if the low-flow showerhead is saving water even though I'm taking longer showers now. If they aren't, the cost of that electricity and water comes out of my pocket. When I have skin in the game, the cost of failure is paid by me. This isn't always the case with political decisions. Few politicians are environmental experts, or scientists, or economists, but we ask them to make decisions requiring all these skills. They are experts, however, in what the public wants, so they often choose environmental fads—policies that sound good but may do little for, or may even harm, the environment. They make decisions based on what they know, not necessarily what is effective. What sounds cool isn't always what helps the environment.

The results can be extremely bad. For example, I spent years researching schools built to politically mandated "green" building standards, from North Carolina to Washington State. Consistently, in state after state, the so-called "green" schools used *more* energy per square foot than traditionally built schools in the same district.[11] Facilities directors at more than one school district told me in confidence the motivation for using the standards was that the school board wanted to show they care about the environment, and a plaque on the wall provided tangible evidence of their environmental conscience, even if the monthly electric bills didn't. When politicians are presented with data showing their decisions may actually be harming the environment, they often ignore the information, claiming it is wrong. The last thing politicians want to do is admit they made a mistake and face public embarrassment. When was the last time you got someone to admit they were wrong on Facebook, in front of their friends and the world? The incentive for politicians to defend their image as an environmental champion is at odds with the real-world data showing the policies are polluting the environment. As the *New York Times* columnist Bret Stephens noted when discussing the many similar failures of climate policy, "We need to make policy choices based less on moral self-regard and more on attention to real-world results."[12]

The information available on our smartphones helps individuals align personal incentives with the environment—making people pay the price when they waste resources—and encourages them to correct mistakes. Unlike politicians, if I realize I made a mistake, I can quietly correct it without having to face the judgment of friends and neighbors.

Helping individuals to make better decisions or generate more accurate information still doesn't ensure those actions add up to meaningful improvement. Small, environmental technologies turn individual efforts into a wave of change. Anyone who doubts this can look to the way Uber has radically changed the taxi industry simply by connecting two people—a rider and a driver—millions of times. Neither rider nor driver intended to change the taxi system as a whole, but the collective efforts of those who find each other simply by clicking on their phone changed an entire industry in just a few years.

The opportunity for targeted technology solutions has also expanded dramatically. Rather than needing a major donor or corporation to back an invention, innovators can focus on solutions that make a significant difference in a specific area. Such solutions don't need lobbyists to mandate their technology or receive taxpayer support. Instead, a few people with a good idea can create a smartphone app or build a technology that solves a particular problem. There are almost daily stories of major new environmental technologies. Simply search for "recycling ocean plastic," and you will find hundreds of articles highlighting technologies that turn plastic waste into fuel, new plastic, shoes, even beach huts.[13] These innovations often require massive investments and large companies behind them. By way of contrast, the Seabin, which collects trash in marinas, was created by a couple of surfers who grew tired of seeing garbage in the water where they surfed. Their technology was backed by an incredible crowdfunding campaign with donations as small as two dollars. Seabin, and other small technologies, may not have as big an impact as other efforts, but they solve problems that would otherwise be ignored because they are too small to receive the attention of major investors or politicians.

This approach transcends politics. In the examples shared in this book, we will find people as politically diverse as the former mayor of proudly progressive Santa Cruz, California (where the university mascot is the banana slug), farmers in Alabama, and young computer programmers from Jakarta. Each of them is using small technology to help the environment in different ways: helping fish and fisheries by improving water quality, reducing the amount of water we waste, and improving the management of ocean fisheries. Their backgrounds and politics could scarcely be more different, but thanks to technology their efforts converge, with each of them playing a role in solving a much larger problem. Aligning politics is increasingly difficult. Political trends can be erratic. Laws can be repealed. It is almost impossible, on the other hand, to reverse technological innovation, especially when it saves money. Everyone can appreciate that, no matter where you live or who you voted for.

For the past several decades, the message of too many environmental activists has been that we need to dramatically change our society and economy. One activist at a Seattle-area environmental group told a reporter that her goal was "remaking the economy of the nation, the whole globe."[14] Often the focus is on forcing people to change their behavior. Doing that, however, is very hard and is likely to end up causing unintended consequences. When these grand approaches fail, the result is cynicism, as people may resign themselves to environmental degradation.

The visionary innovator Buckminster Fuller had a different approach. "You never change things by fighting the existing reality," he wrote. "To change something, build a new model that makes the existing model obsolete." That is the spirit of cutting-edge environmental technologies. Innovators across the globe are using the power of the technology in their hand to change the environmental reality on the ground. They are expanding access to renewable energy, saving endangered wildlife, cleaning the air and water, and protecting honeybees.

People are already making wonderful environmental progress, building a new model that is changing how we act as stewards of our planet.

ACHIEVING WHAT POLITICIANS AND BUREAUCRACY CANNOT

We now have the option to choose to work together through smartphones, creating instant communities and collaborating in new ways. Rather than replacing existing government solutions, personal technology targeted to serve the goals of one of these instant communities fills in where government falls short. Sometimes government approaches simply don't work or aren't needed. In Nicaragua, the presence of hundreds of soldiers with AK-47s on the beaches isn't stopping the poaching that threatens sea turtle populations. In California, the Nature Conservancy discovered that a collaborative approach to providing bird habitat could be exceptionally effective and relatively inexpensive. They didn't have to spend years lobbying for rules that may or may not be effectively enforced by underfunded government agencies.

Where there is politics, there is controversy. Climate change has become one of the most divisive issues in American politics, which Gallup Poll Editor-in-Chief Frank Newport called, "one of the most remarkable examples in political polarization of the past two decades."[15] Those who were happy with the environmental policies of Barack Obama despaired as Donald Trump moved in a different direction, and vice versa. If solutions rely on bipartisanship, there aren't likely to be many successful environmental policies enacted in the near future. The fights seem to be getting nastier. The divide between the parties is growing. Taxpayers feel more squeezed for resources. As economist Tyler Cowen notes, "It's hard to buy off the various interest groups because government revenue is down, and they become more and more likely to engage in a 'fight to the death' over political control."[16] One result is that environmental policy swings wildly as political power moves between factions.

Politicians, fearing the political price of controversy, avoid hard choices. In the Puget Sound area, studies show that the ravenous and growing population of seals and sea lions plays a significant role in the decline of salmon populations,[17] while oil tankers have had almost no impact on the fish. Yet politicians have repeatedly tightened regulations on oil tankers and increased funding for tug escorts. With regard to the charismatic marine mammals, politicians are virtually silent. Certainly, a major oil spill would be significant, but there is no disputing it is easier to focus on a theoretical threat from Big Oil than the everyday harm being done by furry sea creatures.

Some readers will consider me too alarmist regarding climate change or other environmental concerns. Others will feel I am downplaying some risks. My purpose is to highlight the growing number of opportunities to make those disagreements less important. You may think I am not worried enough about climate change, but we can both cheer the innovators who are making it easy and inexpensive to reduce CO_2. Others may believe I worry too much about drought and water shortages but will appreciate the innovative technologies that make it easy to reduce water waste and save money. The less environmental action is contingent on political agreement, the more successful we are likely to be.

It is also difficult for government policies to adapt rapidly to the emerging issues of the day. The United States has been talking about a climate policy for more than two decades. Policies that favor one technology or approach remain in place long after they have proven ineffective or been surpassed by emerging technologies. New technologies have difficulty competing, because instead of just a good idea and some money, innovators may need a good idea, money, and a lobbyist to compete with entrenched technologies. Tom Ranken, who ran the CleanTech Alliance in Seattle and was himself a tech innovator, told me, "Innovation may occur in developing countries where impediments are fewer." He's right. Some of the most exciting applications are created in developing countries, where governments simply don't have the ability to address environmental harm. The stability provided by existing rules and bureaucracies can easily become stagnation and a hindrance when disruption is needed.

Real progress will require a change in how we think about environmental solutions. It may seem unlikely that individuals with smartphones will be able to do what the Environmental Protection Agency or the United Nations cannot. But small actions add up to meaningful results. Each targeted innovation makes use of information, incentives, and collaboration in a way that can't be matched by traditional political approaches, which are limited in their ability to innovate and tailor policy to the needs of individuals.

Ultimately, I agree with Louis Pasteur, whom I quoted at the start of this chapter: the impact of the small can be extremely large. As a beekeeper, I know how the work of many small individuals can add up to remarkable results. Each little bee does its small part. They communicate and collaborate in very simple ways, sharing information with others around them. In most cases, the hive runs well. It is a system that has worked for millions of years and has adapted to environments across the planet. It is a model we should emulate.

At a moment when there is so much anxiety and pessimism about the environment, the power of the technology we have in our own hands is a source of real optimism. Our environmental problems are real, but many hands make light work, and with the opportunities created by smartphones and personal technology, they can be solved in ways that were unimaginable just a decade ago.

The opportunity to be an environmental hero is now literally in the palm of our hands.

The Power of Small, Nimble Innovations

Tito Jankowski is worried about climate change and the billions of metric tons of CO_2 humans put into the air every year. He wants to grab some of that CO_2 and bring it back down to earth. Despite the size of the problem, his approach is distinctly minuscule. "People underestimate the power of things that start small." Focusing on incremental progress, he believes, helps prevent becoming demoralized by the size of environmental problems. "I prefer not to get overwhelmed," he explained. "There is a theme in environmental stuff: 'Let's get overwhelmed together.'" Jankowski wants to focus on things he can understand.

To start, Jankowski launched airminers.org, a web community of companies developing technologies that remove CO_2 from the air to reduce the risk from climate change. There are machines—big, expensive machines—that remove CO_2, but what can be done with the material they capture? Some are pumping the CO_2 into the ground to store it. Jankowski wanted to make something from it. "Here are these materials pulled out of the atmosphere; what can we make from those," he wondered. "We made a planter out of carbon and sent it to our e-mail list, and they sold out in an hour."

This seems like small stuff. And it is. Small innovations, however, will play a larger role in solving some of the world's largest environmental and resource problems. This is the honeybee approach to sustainability. To make one pound of honey, bees must visit two million flowers, and the average bee makes only one-twelfth of a teaspoon in her lifetime.[18] Within the hive, there are fifty thousand honeybees, each of whom has a role to play. Some care for the queen. Others protect the hive. Some collect nectar and pollen. When there is a shortage of something, bees switch jobs. If a honeybee stopped to think about it, the job of providing for the hive would be overwhelming. As daunting as it seems, it has worked for millions of years. Make a contribution, no matter how small. Don't get overwhelmed. It works for bees.

Small technologies make the honeybee approach work to solve some of the most daunting environmental problems the world faces. Implementing that theory has been easier said than done. To be effective, small technologies must be adopted by sufficient numbers of people to have a meaningful impact or there must be a large variety of different technologies each playing an important, albeit limited role. The barriers to creating innovations that solve specific environmental problems—playing their role in the larger hive of solutions—are lower than ever, and the ability of creative people like Jankowski to have their inventions go "viral" and reach large numbers of users is growing all the time. The role of the small is becoming increasingly large to solve some of the world's biggest environmental problems, like those included in the United Nations Sustainable Development Goals.

Adopted in 2015, the Sustainable Development Goals (SDGs) set targets for improvement in seventeen areas, including improvements in economic prosperity, health care, education, and the environment. Prominent among the SDGs is access to clean water. Goal number six is to "ensure access to water and sanitation for all."[19] The preamble explains that "due to bad economics or poor infrastructure, millions of people including children die every year from diseases associated with inadequate water supply, sanitation and hygiene." Nearly 800 million people around the world don't have access to clean drinking water.[20] The World Health Organization estimates 485,000 people die each year from illnesses associated with contaminated drinking water.[21] There have been many efforts to address this problem, but the results have been poor. Even embarrassing.

In Egypt, for example, the Ministry of Health used funding from the Agency for International Development to build a water system that delivered chlorinated water to villages. Despite that, the villagers continued to collect potentially contaminated water directly from the Nile. In his description of the project, Everett Rogers, one of the world's leading experts on the spread of technology, noted there were several problems with the approach offered by the government.

First, the system to turn off the spigot was often faulty and easily broken by the users. Rogers explained that "each spigot was originally equipped with a spring-loaded shut-off valve, so that the flow of water would stop when the valve was not held open. However, the constant use of this valve often broke the spring."[22] That suited the people in the village; they preferred to leave the water running. "Many of the springs were intentionally broken by villagers, who preferred constantly running water," wrote Rogers. "So pure water ran out of the spigot day and night, creating a filthy mudhole around the spigot." This wasted huge amounts of decontaminated water, leaving pools that became breeding grounds for mosquitoes.

Villagers were also skeptical of what the government was putting in the water. Some believed the chlorination was intended to reduce the population. Others said the water had a bad taste. And, while gathering along the Nile had a social component—women washing the clothes together and talking—standing in line for water was unpleasant. When the water unexpectedly turned off, there were fights among those who were waiting.

These problems aren't unique to Egypt. Even where there is funding to install hand pumps, about half are broken at any moment because funding for maintenance is not available.[23] In communities where the taps worked, frequently only one person could run the water tap to prevent the type of waste that occurred in Egypt. If someone wanted water, they had to find the person who controlled the water tap. Alex Burton, a former commander in the UK Maritime Forces, became interested in how to address the lack of clean water in African nations. "People were paying for water but weren't getting the reliability," he told me. The frustration of villagers caused tension. "We were most worried about the person running the tap. The person running the tap used to be chased around the village."

Some new water pumps come with internet-connected sensors to determine if they are working and how they are being used. This data is useful but is not by itself enough to address the misaligned incentives and lack of resources that make water managers unresponsive to the needs of communities. Simply knowing a pump is broken does not mean someone has the incentive to fix it. The basic problem is that the livelihood of people who manage the water systems is not related to the delivery of water. If water managers lose money when the water isn't flowing or when villagers don't feel the water is what they want, those overseeing the water system will have incentives to fix it.

That is the theory behind eWATERservices, a system Burton helped develop that uses smartphones and technology to create a financial incentive to keep the water flowing and facilitate both the monitoring of water systems and payment for the water. Called "PayPal for water," eWATERservices customers use a small device like a key fob for a car to open the spigot. The system uses Amazon's cloud services to keep track of payment data and the working status of taps. In their case study of the project, Amazon's blog noted that "eWATER's founders saw a solution in the wide use of cell phones, which are now omnipresent in even the poorest countries. They reasoned that if people were willing to pay small sums for cellular service, they would also be willing to pay small amounts for clean water at the tap, provided that the money collected would go towards reliable maintenance of the water systems."[24] Villagers living near an eWATERservices spigot get a fob attached to an account they have paid into. Each time they use the water, eWATERservices tracks how much water comes from the faucet and debits their account. Users can refill their account by using their cell phone. All of this is automatic. Users can turn the tap on and off by themselves without having to track down the person managing the tap.

The critical element is that the income of water managers is contingent on the taps being functional. Every day a tap is broken represents lost income. The revenue from the tap also provides the resources necessary to fix the taps. Real-time data about the status of each spigot are available online to identify where there are problems and fix them quickly. The process of repair can also be tracked and tested online. One report noted, "Engineers complete jobs and send simple commands to a dispenser from a basic mobile phone to diagnose faults. Once a dispenser is fixed, engineers can complete mobile service logs to ensure continuous improvement of water systems quickly."[25]

The idea of combining real-time tracking with financial incentives was a good theory, but Burton knew the eWATER-services spigots would have to be tested in the communities they hoped to serve. Top-down solutions, like Egypt's effort to provide clean water, are often well-intentioned but fail to address hidden barriers that stand in the way of achieving the goal.

One of the early tests was in Gambia, where some community taps had been locked for most of the time when an attendant was not there. After eWATERservices installed eighteen solar-powered taps, rather than having to wait, people could collect water up to nineteen hours a day.[26] In Tanzania the eWATERservices taps tripled the revenue from each water point and cut waiting times from hours down to about ten minutes.[27] "We then manage the cash, and we hand it straight over to the company," says Burton, "and his revenue is increased by 340 percent."

The flexibility and power of the technology made it possi-ble to solve the problems faced by traditional water projects. It also makes it possible to track the results and adjust to problems. Anyone can go online and see the results for themselves. As of November 2021, eWATERservices had expanded to three African countries, with a total of 518 water dispensers serving more than 175,000 people, 97 percent of which are working as I watch the live data online.[28] Since the project was launched, it has delivered more than 500 million liters of water.

One write-up of the project was effusive about the opportunity created by small technologies, sometimes called the "Internet of Things" or IoT. "It has become clear that the use of IoT is truly changing people's lives by making a sustainable and positive impact on the most critical and basic of human needs. The power of connected technology to solve issues and truly change lives across the globe cannot be underestimated," Burton says. He believes the opportunity is greatest where governments are weak. "I certainly think there is an energy around the use of mobile phones that is leapfrogging the West," he said. "That has been synchronized with the Internet of Things. That has really opened the opportunities for innovation."

The breakthrough wasn't simply adding the technology. The technology to deliver water through pumps has been available for centuries. The key was to match the technology that was newly available with the local circumstances. Since there was no government agency capable of consistently managing and maintaining the water pumps, adding technology to collect payments and run the taps was the missing link. Previous efforts to build water infrastructure failed because they did not address this shortcoming or attempted to address it in ways that were inflexible, such as having a single person control the water tap.

Everett Rogers notes that this is a classic pitfall of top-down efforts to distribute innovation. "Anthropologists often show that the planners and officials in charge of [technology] diffusion programs failed to account fully for the cultural values of the expected adopters of an innovation," he wrote. "As a result, the diffusion program often failed or at least led to unanticipated consequences."[29] Prior to the availability of smartphones and the portable technology to run the eWATERservices system, the communities did the best they could. As harrowing as it was to be the person controlling the water tap, it may have been the best option. That is no longer the case.

A reliable water system is also good for the environment.

First, because users pay for the water they collect, they will be careful to conserve and prevent waste. Alison Wedgwood, who became the full-time CEO of eWATERservices in 2020, explained that "as soon as people are paying for water, they are valuing it, and that means they don't waste a drop."

The certainty that the water is clean reduces other types of environmental harm. When water pumps are not working, people collect well water that is frequently contaminated with worms that spread illness. To prevent that, villagers buy plastic bags of water. They can be sure the water is clean, but the result is a big increase in plastic waste. Working water pumps means fewer plastic bags.

The other way to be sure water is not contaminated is to boil it. "In Tanzania and Kenya," Wedgwood told me, "there are charcoal sellers everywhere and they cut trees down. The landscape is being denuded of trees." Deforestation is a major problem in some parts of Africa as people cut down trees to cook their food and boil water. Clean water from a spigot means people do not have to buy charcoal. That's not only good for forests, it reduces CO_2 emissions. "We have talked to the Climate Trust about how much CO_2 we can save because they are not burning charcoal," said Wedgwood.

The eWATERservices spigot is effective at reducing environmental impact because it empowers individuals and makes their life better. This is a point Wedgwood stresses. "It is a crazy situation when you have kids who are still walking—500m is still considered accessible—carrying 20kg of water, basically a suitcase of water, on their head." Instead, it costs about a penny a day for the same amount of water in the village. "If we make it easy to access water, women can go to work, and girls can go to school."

Empowering people also has another benefit. "We are definitely disruptive," says Wedgwood proudly." The typical approach is for NGOs in developing countries to install a water pump and then turn over control to the "community." Wedgwood sees that system as fundamentally flawed. "The idea that it is good to have village ownership is a myth I want to smash," she says. "The current model sounds good because you hand over the system to the village and they have ownership. There is always a head man. You are giving it to an old man and his mates and cousins. And you have given it to this patriarchal system." When NGOs recognized this, they started to require the committees that controlled the pumps to include women. That, she explained, did not change things. During the committee meetings, "the women are always sitting on the floor and the men are sitting on the chairs." Turning women into customers means the system is more likely to respond to their needs, disrupting the typical power structure.

To address something as big as water access—and the systems that control access—the assumption is that we need big solutions. There is a term for this mentality: "proportionality bias."[30] We tend to believe that big events have big causes. This is one reason people believe conspiracy theories. It doesn't seem plausible that a series of small events could lead to historic outcomes, so we assume that hidden forces are at work. The corollary is also true, that it is difficult to imagine solving big problems without big solutions. But big solutions have been tried. Billions have been spent and access to clean water is still a major problem, with about two thousand children a day dying from diarrhea from unclean water.[31] In fact, the systems that could deliver big solutions have been a source of the problem.

Small, flexible approaches like eWATERservices increase the likelihood of finding effective solutions and spreading those innovations to other communities with similar circumstances. Small solutions are easier to create and test. It is easier to discard those that don't work and develop new ideas that have promise. Small innovations can be copied, maximizing the potential range of application. Others can be scaled up, taking advantage of economies of scale (starting small doesn't mean good ideas should stay that way). Low barriers to innovation expand the ability of people closest to the problem to innovate, so they don't have to turn to major donors like corporations or governments. Research on the history of innovation shows local communities with an interest in conserving resources are the fertile ground that facilitates creativity and progress.

THE RAPID SPREAD OF SMALL INNOVATIONS

Small innovations have several advantages over large-scale, top-down projects.

To begin with, it is difficult to know what technologies will emerge—what will be successful and what will fail. Putting all or many eggs in one basket is extremely risky. In the late 1960s, many believed supersonic transport would be the technology of the future. Despite being touted as an alternative to fossil fuels, ethanol made from corn or palm oil sometimes does more harm than good. Committing large amounts of money to one technology, or using policy to favor a particular approach, is a high-risk strategy. History has not been kind to large innovations imposed from above.

Second, research demonstrates that centralization is associated with stagnation, not innovation. This is true for large corporations or governments. In his work on the creation and diffusion of innovation, Everett Rogers notes, "The more power is concentrated in an organization, the less innovative the organization is." He argues that "in a centralized organization, top leaders are poorly positioned to identify operational-level problems or to suggest relevant innovations to meet these needs." When decision makers are far from the problem, they do not have the necessary information to judge the merits of the wide range of innovations under their authority. Once centralized organizations make a decision, Rogers notes, they may be able to rapidly implement new approaches. That can be a double-edged sword if they select the wrong innovation. Massively and quickly implementing a bad idea can make the problem worse.

Rogers also found that individuals adopt new technologies based primarily on the recommendation of friends and trusted colleagues. If a neighbor uses a new technology and likes it, that is far more likely to be definitive in deciding whether to adopt it than stories in mass media or the recommendation of distant and unknown experts. While mass media can be effective at raising awareness of new technologies, the recommendation of a neighbor is more important to the final decision. Even if governments or large businesses want to impose a new technology on a mass scale, the target audience is unlikely to go along without the approval of friends. Those who must live with the costs and consequences of new innovations may also perceive that governments have different interests than they do. When police departments across the United States partnered with Amazon to distribute Ring doorbell cameras to homeowners, there was backlash when homeowners and community members realized the video from those cameras could be accessed directly by police.[32] Many people who would immediately share the video from their camera with the police, or even on Facebook, were nervous when that same technology came directly from the police. People place a value on having control of the technologies they use and were willing to forgo the subsidy to buy the Ring in order to have complete control of the video.

All of these limitations on diffusion of small, personal technologies—the need for interpersonal trust and a reluctance to adopt until others have—would seem to imply that it takes more time for them to have a meaningful impact. The reality can be the opposite. When the spread of technology is based on user choice, the impact can be felt very quickly. Rather than having to spend money imposing a technology and risk political backlash, a decentralized approach engages the most powerful force in innovation distribution: the trust of local experts and neighbors. Small technologies are more likely to go viral and spread rapidly. As Rogers notes, "Innovations requiring an individual-options innovation-decision are generally adopted more rapidly than when an innovation is adopted by an organization. The more persons involved in making an innovation-decision, the slower the rate of adoption."[33] In large organizations—governments or business—many people must agree.

There is also a fundamental difference between the argument made by governments looking to impose technologies and the decision-making process utilized by individuals when considering new techniques or equipment. When farmers, for example, consider a new piece of equipment, they focus on the visible benefits being enjoyed by their peers. The results will differ slightly, but if someone else is enjoying increased crop yields or reduced costs, chances are they will too.

Technologies imposed from the top make a different argument. Trust us, they say; this will be better for you. They may have data or other indications that technologies would be advantageous, but at the end of the day, the very fact of "imposing" a technology implies the benefits to the user are limited or speculative. This doesn't mean imposing technologies is always inappropriate. It does mean that there are limits.

The lack of tangible benefits explains why some innovations have been adopted ahead of more meaningful improvements. Doctors began using anesthetic quickly because the benefits of operating on an unconscious patient were immediately obvious. It took decades, however, to convince doctors to wash their hands, because the benefits were not obvious. Connecting a lack of hand washing to illnesses suffered by their patients was difficult, and doctors were skeptical until germ theory was understood, making the connection tangible.[34]

The same is true for hard-to-measure environmental risks like climate change. Partisans tend to assume that the more people know about the science of climate change, the more others will agree with them. In fact, the research shows that the more people know about the science, the more divided they become—moving to the extremes of the issue—using selective information to reinforce their own biases.[35] Research in the *Proceedings of the National Academy of Sciences* found that "individuals with greater education, science education, and science literacy display more polarized beliefs" when it comes to climate change, evolution, and other controversial science issues. The intangibility of the impacts of climate change leaves people to apply their own beliefs and level of risk tolerance. During the COVID-19 crisis, some wondered why so many people who questioned the seriousness of climate change were careful to self-quarantine and maintain social distancing. As the death toll mounted it was easy to see tangible everyday evidence of the risk. By way of contrast, even those who believe climate change is a risk differ widely in their assessment of how serious it will be.[36] The promise that a technology will help reduce the impacts of climate change fifty years from now will appeal to many fewer people than the ability of a CO_2-reducing technology to save money today by improving gas mileage.

Ironically, if we want to address environmental problems of vast scale, it works better to find small approaches that are personal and tangible. Rather than relying on discovering a winning political argument, offering immediate benefits takes divisive and subjective factors out of the equation. Small innovations offer progress people can see, which helps build momentum.

For example, Rogers tells the story of homeowners in the Dominican Republic who purchased solar panels for their homes. Frustrated by the lack of responsiveness and service from the government-run electric company, homeowners who were too often sitting in the dark looked for an alternative. Rather than looking to overthrow the existing utility or restructure the entire system, they made a small decision with tangible benefits. When the government-run electric company began offering low prices to homeowners to discourage the purchase of solar panels, many people decided the real improvements today outweighed the intangible promises about a theoretical future. "People soon learned that photovoltaic electricity was better, once they experienced blackouts and other problems with the power utility," Rogers wrote. "PV [photovoltaic] never goes out. Plus, we own it," said one happy homeowner.[37] They didn't have to be convinced of potential environmental benefits. The primary motivation was the immediate benefit solar panels offered to the people living in the house: reliable electricity.

Clean water and climate change aren't the only issues we can address using small, tangible technologies. The increase of plastic in the ocean has also become a growing environmental concern. For two surfers who had to navigate trash in the water, the problem was immediate. Rather than solve the problem of ocean plastic, they decided to address a problem they saw every day.

HOW TWO SURFERS ARE CLEANING THE OCEAN

Pete Ceglinski was tired of trash in the water where he surfed, so he decided to do something about it. "I'd been mucking around for three or four years and then decided to get serious in the last two years," he told me in 2018. "I quit my job, found a factory, and started a crowdfunding campaign."

Pete and his business partner, Andrew Turton, called their creation Seabin, and today it is cleaning marinas as distant as the Caspian Sea off the coast of Iran or San Diego or islands in the South Pacific. It began very simply as a crowdfunding campaign and ended up raising hundreds of thousands of dollars from eight thousand people.

As surfers, Pete and Andrew realized that trash from local marinas was drifting into the ocean. Plastic and other items would fall off the moored boats and make their way into the waves where Pete was surfing. If there was a way to capture this trash from marinas before it reached the ocean, it wouldn't solve the problem of ocean plastic, but it would make a difference.

The initial concept was straightforward. There are trash cans on the ground; why not in a marina? Ceglinski and Turton developed a technology that would work in the same way—a bin that would float in a marina and collect the trash automatically. Attached to the bottom of the bucket is a small pump that draws water and trash into a mesh bag that can be changed. Seabin can also filter out oil and other floating pollutants, returning the water through the pump and into the sea cleaner than it came in. The goal was to sell it to marinas and yacht clubs, but they needed to prove the concept first. To do that they needed funding.

That's when they turned to the crowdfunding site Indiegogo. Initially, Indiegogo was created to fund independent movies. The model worked well, so they branched out to supporting musicians and creative campaigns. By the time Seabin came along in 2016, crowdfunding had moved into helping innovators create new products and technology in a wide range of fields.

Fundraising was a dilemma for Seabin. On the one hand, the marinas and yacht clubs they hoped would use the product were unlikely to buy one without some promise that the product was affordable and would work. The word of two guys who were tired of trash in the surf wasn't going to be enough. On the other hand, the standard crowdfunding model—where people pledge to buy a product when it is produced—might be difficult because Seabin wasn't intended to be a product an individual could use. They had to hope crowdfunding would connect them to people who might give only a few dollars but wanted to clean the ocean.

"In the end it was absolutely amazing. People would donate three dollars or five dollars out of their pocket money," said Ceglinski. "It was really humbling." They raised $267,000 from eight thousand people.[38] People who would never see a Seabin were contributing, providing the resources to begin the project development and the research. One of the first things they did was develop prototypes to work out the kinks. After testing, they had a good idea of what kind of trash they would encounter. When I spoke with Ceglinski, they were still in the testing phase. "We've spent nine months collecting data from six locations around the world. Last year in Montenegro, one of the Seabins caught 876 kilos in six months. They are getting about a ton of debris per Seabin." Much of what they've found in the bins is relatively small. "The most common item is cigarette butts and microplastics and plastic particles and food wrappers," explained Ceglinski. "Another thing we see every single time is polystyrene—they are everywhere."

All of this would have been much more difficult without crowdfunding. Although many angel investors are charitable, they typically invest in projects that have a good chance to achieve success and return several times their initial investment. Projects like Seabin are unlikely to have that kind of success. There are some who are trying to change that. Brock Mansfield launched the Salmon Innovation Fund to connect investors to those working to rebuild populations of Pacific salmon and steelhead in the western United States. Despite the environmental focus, the fund still has its eye on the bottom line, noting they expect the innovators they back to "exit outside of the fisheries industry, allowing for competitive returns to our investors."[39] That's a good thing, and the Salmon Innovation Fund has funded a few projects that help promote sustainable fisheries, including a product verification system to track the seafood supply chain and ensure it was caught legally and in a sustainable manner. Finding projects that are financially rewarding is fantastic, but not every project can attract major investors. Crowdfunding expands the opportunity for projects that may not be ready for large, cause-focused investors.

In addition to reaching a broader base of potential funders, Ceglinski found that crowdfunding served to promote his product to potential buyers. "Crowdfunding is an amazing platform. Even if you can't raise your target funds, your idea has an incredible global platform. It is like a free marketing tool." With the attention earned from the crowdfunding effort, Seabin was able to raise additional funds and find clients for the product, including Safe Harbor Marinas, which manages dozens of marinas in the United States.[40] Despite the positive results, Ceglinski admits it was difficult. "It is a stressful process and takes all your energy." But it gave them an opportunity to raise money that would not otherwise have been available. "If I needed to start from zero," he says, "I would use crowdfunding."

By fall of 2021, there were more than 860 Seabins across the globe, collecting about four tons of trash every day.[41] "There are lots of small differences that everybody can make," says Ceglinski, and crowdfunding is creating new opportunities to do that.

The key role that crowdfunding plays is to help small innovators reduce the cost of developing prototypes and refining a solution. With crowdfunding, the risks associated with innovation are much smaller than creating technology that requires a large customer base. Innovators must still create a solution that appeals to their audience—big or small—but the barriers are lower. Crowdfunding companies like Indiegogo and Kickstarter now have staff who are dedicated to helping environmental innovators make those connections.

Enzo Njoo is head of sales at Indiegogo, and he has a personal motive for helping environmental innovators. "Born in Indonesia, environmental activism is a big part of my passion and life. I grew up as a scuba diver because Indonesia is made up of seventeen thousand islands," he said. "I went back to where I grew up and you can see the impact of environmental damage," he told me. "You can see the impact to coral species and the food pyramid. Deforestation in Borneo with palm oil plantations taking over." Njoo uses those memories as motivation to help sustainable products get their start and build momentum. "The tide has already been turning towards sustainability and green technology," notes Njoo. "What I'm hoping to see is not just environmental sustainability being a cause but being the actual main product."

Indiegogo's environmental crowdfunding program hopes to be the tip of the spear of environmental innovation. Crowdfunding provides opportunities for niche products that solve a small problem. It can also be used to refine a prototype before reaching a larger market. The ability to start small and grow catalyzes the change Njoo has in mind. "Crowdfunding is useful because we de-risk the market validation," he said. "It helps you understand whether what you are doing—product or cause—whether there is any demand for it. Whether other people care about it before you have to raise too much funding. It is the way to test your assumption and fine-tune it." Limiting the cost of failure avoids false starts that can undermine corporate and consumer support for green products.

For example, Tully's Coffee had to back away from their compostable cup after they found the cup leaked when hot. The coffee company announced, "Due to leakage issues with the Ecotainer cups that we were unable to resolve, we had to switch to a different cup, which is made from 10 percent post-consumer fiber."[42] Product testing can avoid embarrassing failures for tinkerers as well as big business.

Much of the support for crowdfunded projects comes from early adopters—people with disposable income who care a great deal about a cause and are willing to spend their money to support it. They are willing to take risks on innovative products because, if the products work, they want to be able to say they were the first to own it. For some crowdfunding projects, small donors receive a certificate that tells the world you were one of the visionaries who helped launch a great idea. People giving fifteen dollars to the Seabin campaign received a certificate of appreciation, which is pretty cool now that the project is having success. By targeting that key group of early adopters, small innovators don't have to develop something that must sell to a million people.

Early adopters are the connecting tissue between innovators and a larger market audience. They have the disposable income to try out new technologies and determine if they work. These early adopters are often respected voices and, as Rogers explained, "serve as a role model for many other members of a social system. Early adopters help trigger the critical mass when they adopt an innovation."[43]

How innovators find early adopters to trigger that chain reaction of technology diffusion isn't always straightforward. For the agricultural technologies Rogers studied, the early adopters were community leaders sought out by innovators or change agents "as a local missionary for speeding the diffusion process." When the community is virtual it isn't clear who those missionaries might be or how to find them. Crowdfunding sites create that online community where early adopters connect with innovators. Crowdfunding is a like a farmers' market for innovation—where buyers and sellers of unique products go to find each other. In creating Seabin, Ceglinski and Turton found that crowdfunding provided more than just the resources to launch their product. It also built a community of early adopters who promoted the idea to other potential buyers.

Another crowdfunding site, Kickstarter, recognizes their role is not just to reduce the cost of innovation but to help early adopters find technologies they support. When they launched a program called Shapeshift to encourage innovators to use recycled materials, they also made a commitment to drive early adopters to those environmentally friendly products.

Kickstarter launched Shapeshift by publishing a guide to environmental design, developed with the Environmental Defense Fund (EDF).[44] The guide helps designers think about durability, repairs, how the product can be recycled, the packaging, and other aspects that may not be top of mind even for environmental innovators. The campaign highlights products made "from recycled plastics, papers, textiles, ceramics, coffee grounds, fishing nets, air pollution, and more."[45]

I spoke with Kickstarter's VP of insights and environmental impact Jon Leland, who led the effort to make the organization, and products on their platform, more environmentally friendly. He says Shapeshift is a method of "highlighting this way of building a product with sustainable materials—driving backers to those products through promotion. This is a perfect opportunity for people to look at the existing mode of production and find ways to make it more sustainable and disrupt unsustainable products."

Before joining Kickstarter, Leland worked with several environmental nonprofit groups, including the EDF and the Natural Resources Defense Council. He moved to Kickstarter because he believes we need to tackle environmental problems from many angles, and using small, crowdfunded projects can make big changes. "Crowdfunding is able to accelerate change" and disrupt large businesses that don't have the same entrepreneurial opportunity. Leland believes "government needs to play the leading role, although they are failing." I think he underestimates the value of the work he is doing and places too much faith in politicians. Government has a role to play, but technological solutions are happening faster than government can keep up—which is great, because it helps accelerate the pace at which we can address environmental problems.

Rather than having government spend taxpayer money on trendy, but ineffective or costly, projects, crowdfunding uses discretionary income from early adopters who want to be part of environmental change. And if they fail—as many will—innovators learn a bit more about what might work better, and those who lose money do so with their eyes open.

Tito Jankowski, whose small-ball approach I mentioned at the beginning of the chapter, is one of Shapeshift's innovators. After moving to Silicon Valley and working in biotech, he started to think about climate change and what he could do. He became intrigued by technologies that could remove carbon from the atmosphere, a technique known as carbon sequestration. This is what inspired the development of his planter made from carbon collected from the atmosphere.

He put his next project on Kickstarter as part of the Shapeshift campaign. Called "Negative," it is a black bracelet made with carbon that was once in the air. "People had heard about these technologies, but they had never seen anything, and this is part of the future they wanted to see," he said. "It would be better if they could have something they could carry around with them." He launched the bracelet for a price of seventy-five dollars, with a goal of raising ten thousand dollars. It took one day to meet the goal.

For Jankowski, the key to crowdfunding is the community it builds. "We have two hundred people"—many more since I spoke with him—"who have raised their hand and said, 'I want to see this exist.' That is really powerful." That group of interested investors is something he can build on. "I think that hand raising is indicative there are other people. That there is more interest and effort and excitement."

Bracelets and flowerpots are not going to save the world. They do, however, provide an additional incentive to develop carbon capture and storage (CCS) technology. It can cost more than a thousand dollars to capture and store one metric ton of CO_2. Compare that to a price of about ten dollars per metric ton for projects that capture methane—a potent greenhouse gas—from landfills. CCS is still an extremely expensive way to reduce greenhouse gas emissions, but crowdfunding finds people who are willing to take a risk on the technology and add to the diversity of environmental options, funding innovation that can drive the cost down and make it a viable option in the future.

THE CRITICAL ROLE OF ENVIRONMENTAL INNOVATION

Small, but effective, innovations like eWATERservices, Seabin, and Negative are the types of solutions that will be increasingly critical to make progress on a range of environmental problems. There is the impulse to address large problems like ocean plastic with large solutions—big technology, big data, or big government. There is a role for those approaches, but big efforts bring big challenges.

Sheida Sahandy led the Puget Sound Partnership, an agency in Washington State charged with cleaning up the sound. After more than four years leading the partnership, she turned her attention to environmental innovation. "We don't have enough support systems for innovation when it comes to solutions to water-related problems," she argues. Driving innovation, she believes, is a proactive and enabling way to make progress on the environmental concerns that have been the focus of her career. Innovation results in a multiplier effect of progress, she believes. "People focus on doing more and going bigger while doing the same thing they have done before, but is that the most efficient solution? You have to constantly evaluate whether what you are doing is truly taking you in the direction you want to be headed, in the time you have, with the resources you have. If not, you must innovate."

To some ears, efficiency sounds like the kind of economic bean counting that works against environmental solutions. What makes personal technology and crowdfunding so effective, though, is that they are built around efficiently using resources to solve a problem. Rather than making huge investments in a speculative technology, crowdfunding allows innovators to refine ideas and ensure they get the most bang for the buck. Income and taxes are limited resources. Efficiency doesn't mean the environmental challenge is any smaller—it means the strategies to address them are more effective.

Technology opens up new opportunities to solve some of the problems that frustrated Sahandy. To be sure, she isn't arguing against the need for some government solutions, but to supplement them. And she is excited about the opportunity to fund technology that helps, in her words, to "democratize and distribute the solutions, now that we have such distributed problems" as water pollution.

The EDF agrees. In addition to engaging in public campaigns, similar to other environmental organizations, EDF has made a name for itself developing technology. Sometimes that means working with oil companies to identify methane leaks that impact the environment and waste resources.[46] Aileen Nowlan led EDF's efforts to partner with business and develop new environmental technologies. She calls technology the "fourth wave of environmentalism." Environmentalism began with conservation and protecting valuable land. The second wave, she argues, was the series of environmental regulations put in place during the 1960s and '70s. The third wave was the work EDF helped pioneer, working with business to find new environmental solutions. "The fourth wave," Nowlan says, "is based on the rise of innovation and technology and a transformational shift in the speed at which we can get things done."

Developing new technology has become an important part of what EDF does now. "If you started to delve into the specifics on the challenges we face, whenever the technology doesn't exist to solve an environmental challenge, that's where we think about catalyzing that solution." EDF has an established history of promoting technology-based solutions. In 2011 they partnered with Innocentive, a program that connects innovators with those who have different problems that haven't been solved in-house. The goal was to create opportunities for people to profit by solving environmental problems. Beth Trask, who led the program at EDF, recognized that democratizing innovation made sense because "there are many more ideas and possible solutions out there in the world than any given company or organization can tap into on its own."[47] The crowdsourced approach—where people with problems provide an incentive to find a solution—is just the flip side of crowdfunding. Connecting those who care about the environment with innovators who can develop solutions has created new opportunities.

Focusing on technology as a solution to environmental problems isn't new. What is dramatically new is the ability to develop, fund, and distribute small technologies that target local or specific environmental problems and can evolve rapidly to improve their effectiveness.

The ability to think small opens up new, and clever, opportunities. Rather than focusing on the next generation of nuclear power technology, people can develop small power technologies, like a mug that can recharge a cell phone, which is exactly what a group of students at the University of Washington did.

"We started at the University of Washington with three engineers and won the environmental innovation challenge," says Marene Wiley, creator of the JikoPower Spark mug, which created electricity through a thermoelectric exchange. The mother of one of her colleagues had been an aid worker in Haiti and noticed that while many people had cell phones, few had electricity. "Something like 80 percent have mobile phones but only 20 percent with access to electricity. It is a huge disparity that is common all over the world." Marene and her friends decided to solve this problem and came up with a mug that can charge a cell phone. In their Kickstarter campaign, they explained, "The Spark uses a thermoelectric module to convert heat into electricity. The greater the difference between the hot side and cold side of the module, the greater the electrical output."[48] By filling the mug with water and putting it over a fire—the name jiko means "stove" in Swahili—it creates enough electricity to charge a phone.

The target audience for the crowdfunding campaign included campers who wanted to charge their phones in the outdoors as well as those who wanted to fund innovation for people in developing countries. "In Kenya, people will travel miles and pay exorbitant fees—one fourth of their income to heat their homes with kerosene and one-eighth of their income to charge their cell phone," explained Wiley. "If we can get this device, which converts heat from a fire or stove into electricity, and spend less than a hundred dollars, they can pay it off in three to five months." That was the motivation for the JikoPower team.

They started by entering technology competitions and raised enough to travel to Kenya and test their product, doing some small-scale market research. After those initial tests, they turned to Kickstarter to fund a larger project. "We realized we wouldn't reach people in Africa. Most Africans didn't have credit cards and shipping was difficult, so we targeted campers and preppers and philanthropists. We raised $50,000, which was our target." Some of the funding they received was simply charity, designed to help create a technology that could improve the lives of people in Africa and elsewhere.

The crowdfunding revenue was critical because finding institutional investors was difficult. For an investor, a small mug designed for developing countries was a hard sell. Kickstarter offered a way to reach people whose motivations steered more toward the philanthropic than the financial. This isn't to say financial viability isn't important; it certainly was to Wiley and her team. "We were working 16-hour days." They wanted to make this profitable and Kickstarter offered a way to earn some money and prove the concept.

Those who bought the JikoPower Spark are a little better off, creating electricity from the fire they already had to cook their food. It isn't a next-generation nuclear reactor, but it doesn't need to be. And no matter the promise of next generation nuclear, or fusion, or other massively scalable technologies, the JikoPower mug can be created, funded, and delivered before the planning and permitting of big technologies can even finish.

Ultimately, JikoPower didn't take off. They met their crowdfunding goal, sold the inventory they had, and didn't turn it into a viable business. Failure is part of the process, and failure is part of finding the best solutions and discarding those that don't work for one reason or another. A key part of crowdfunding is to set meaningful goals and risk failure. By reducing the cost of failure, however, crowdfunding increases the amount of innovative risk-taking that can occur. This is critically important. In the effort to find environmental solutions, the more people who are looking, the more likely someone is to find something that works. Many—maybe even most—of the ideas will fail. Rather than a major company risking millions of dollars on an idea, crowdfunding limits the risk and potential losses.

Encouraging and learning from failure is an important contrast with political approaches. After spending millions on a government program, politicians are unlikely to admit they made a mistake. It is more likely they will claim the idea simply doesn't have enough money, throwing good money after bad.

Some small innovations will have success and learn enough to take the next evolutionary step to improve their product. Seabin is different today than when they began crowdfunding. As they noted on their web page in 2019, "The team at Seabin Project are working to have the technology in constant evolution and are using every opportunity to further develop its debris recovering capabilities, for cleaner oceans."[49]

A HIVE OF SMALL SOLUTIONS

Small technologies are more suited to the innovative dynamism we need to address changing and complex threats to the environment. When looking for metaphors to guide our thinking about environmental solutions, a hive is superior to a moon shot—many small solutions as part of a larger effort to tackle a problem. In a beehive there is no plan, no director. Each worker has a specific job based on the needs of the hive.

Small is nimble. Small is the ally of creativity. Small can spread rapidly. There is a reason we say something is "going viral" when it is suddenly everywhere. Small moves fast. Small can do big things.

Elephants know small can have a big impact. They are afraid of bees. Conservationists looking to protect endangered elephants are using that fact to protect the majestic, endangered animals.[50] In Africa, beehives are used to keep elephants away from farms, helping them avoid angry farmers who might kill them to protect their crops. What small insects can do to the largest land animal, a hive of small environmental technologies can do to fight pollution and conserve resources.

With new tools like crowdfunding and the power of small technologies, the barriers to creating small, nimble solutions to environmental problems are falling.

Making Environmentalism Pay (for You and the Planet)

As I type this, I can see on my smartphone that my house is using 1,675 watts, and that my dryer and kitchen refrigerator are on, even though I am sitting twenty miles away from home. Entering a rough estimate of my electricity rates into the same app tells me that I am spending about seventeen cents an hour on electricity. Later, when the dryer's heating element kicked on, that amount jumped to seventy-one cents an hour.

I know this because I attached an orange box called the Sense electricity monitor to the wires coming into my house. It connects with my smartphone so I can receive electricity data instantaneously, no matter where I am. One of the biggest electrical draws, as I look at my iPhone, is the bank of incandescent lights in my bathroom. Clearly, I need to swap those for more efficient LED bulbs. The ability to tell what is on in my house no matter where I am in the world makes my wife feel like I am monitoring her at times, but it has also saved us money. When I noticed that our kitchen lights were using about 600 watts, I made a trip to Home Depot and replaced them with LED bulbs that, in my opinion, provide better light. They also cut my energy use from 600 watts to about 100 watts. Even in

Washington State, where we have some of the lowest-cost electricity in North America, it will take only six months to pay back the cost. The Sense device and the app identify where I use electricity over the course of a week, a month, and a year. During the last month, something as mundane as my bathroom lights accounted for about five percent of my total use, more than my refrigerator.

The Sense is a prime example of the way smartphone-enabled personal technologies provide incentives to make environmentally friendly decisions, using artificial intelligence to turn a confusing barrage of data into clear and usable guidance. In some instances, as we shall see, artificial intelligence can even act on your behalf to save money while reducing environmental impact.

What I found using the Sense monitor surprised me, even though I'm an energy geek and have spoken with many architects, engineers, utility employees, and building managers about ways to save electricity. Sense allowed me to receive real data about where I use energy so I could adjust my behavior and upgrade the lighting in my home. This is exactly what Mike Phillips, the cofounder of Sense, was hoping for. "If we can know in detail what is happening in homes, and engage consumers, we should make them be more efficient," he told me. Phillips believes that simple information—where we are using energy and how much appliances cost us to use—can lead to real energy savings. People can reduce their electricity use by up to 50 percent by making some fairly straightforward changes, he argues. "Data is so important because it isn't just one thing. Homeowners all have one or two things where they can say, 'Oh, that's stupid,' and fix it," says Phillips.

The technology to provide that data, however, is very new. Sense does two things that make it a powerful conservation tool. As we've already noted, it delivers instantaneous information about how homes use electricity, providing guidance about useful energy efficiency investments. Swapping lightbulbs is pretty simple. Replacing my air conditioner takes a bit more calculation. With the data from Sense, I can do the math and see if it is worthwhile.

Sense uses sophisticated artificial intelligence and pattern recognition to figure out what appliances I have in my home and when I am using them. Each appliance has a unique signature when it starts and uses electricity. Over time, Sense learns to recognize those signatures and identify appliances. "We are sampling power a million samples a second, and it gives us a super-detailed view of power," explains Phillips. This isn't as simple as it might seem. "The system does have to learn about your home. It is not just knowing what your toaster looks like, but also knowing there aren't other things that might be confused with your toaster," he noted. "The problem is the system does have to do this fairly detailed analysis before we can sort this out well enough." The system is constantly learning, too. Because it is connected to the internet, Sense can update its algorithm and become more accurate. That accuracy opens up new opportunities. Not only can it help your home become more energy efficient, it can also help tune existing appliances. Phillips hopes they will be able to tell consumers if something is about to break down.

Although Sense is an effective conservation tool with existing electricity rates, it can really become powerful if people start to receive real-time price signals related to the current price of electricity. If consumers know what appliances use the most electricity, they can adjust their schedule around peak prices. Sense can also sync with connected devices like lightbulbs, allowing users to turn them off from their phone. This will become important as electrical generation moves to more intermittent, renewable sources like solar, and more people have electric vehicles. If people mindlessly plug their electric vehicle in at 6:00 p.m. when they get home, they are adding electrical demand right in the middle of the peak hours when electricity is expensive and most likely to be generated by coal or natural gas. This is expensive and unnecessary since most people only care that their vehicle is ready when they need to commute the next morning. Smartphones and personal technology like Sense will become valuable as renewables and electric vehicles become more prevalent.

The Texas energy crisis in February 2021 offered a case study of how Sense can help cut costs when prices skyrocket. As severe cold settled over Texas, a combination of factors, including reduced energy supply and increased demand for home heating, drove prices upward. Customers using time-of-day pricing saw their costs jump dramatically. Without reliable information about energy use, some homeowners saw bills of more than four hundred dollars a day. Using the "Time of Use billing" tool on the Sense app, a customer identified only as "John F." on Sense's blog began tracking his electricity use and costs.[51] "No one knows how to read their meter, but you need to have visibility into your electricity usage to feel confident the bill is right," John said. He quickly realized that he should shut down his computer and refrain from charging his electric car. When he did get his bill, he noticed that he was charged for very high energy use at the time when he had shut everything down. Using the data saved by his Sense meter he was able to show that the utility's meter readings were incorrect, resulting in a rebate of $837 from his energy company. Sense quoted him saying, "Had I not utilized conservation to the extreme, and gone on blissfully unaware, I would have a ~$2400 bill."

The energy crisis in Texas is an extreme example, but it demonstrates how powerful information can be in the hands of consumers when it matters most. Providing price signals to utility customers and an incentive to conserve could be a critical part of avoiding future disruptions. Even shaving a few percentage points off demand can make a difference.

These personal technology–based solutions are effective because they put power in the hands of people who pay the costs for failure and receive the benefits of successful conservation. "The solution is not to have complicated pricing that people have to work out," says Phillips. "The solution is not to have the utility decide. It really needs to be consumer-facing automation that allows the consumer to say, 'I want my car to go sixty miles tomorrow.'"

Utilities have some incentive to help conserve. They are legally required to hit conservation targets. But their incentives are mixed. The less energy they sell, the less they earn. Politicians also have a range of conflicting incentives, some of which are based in ideology, special-interest politics, and political image. Consumers, however, have skin in the game and feel the cost of their own decisions, good or bad. Politicians may tell me that triple-paned windows or solar panels will save money, but if those things turn out to be incorrect, I pay the price, not politicians.

Effectiveness requires that we track our success and learn from our mistakes. The more policy makers, homeowners, or others separate themselves from the consequences of their actions, the more likely they are to make mistakes and the less likely they are to address their failures.

Spending money effectively is important because there are trade-offs—every dollar wasted is lost to other causes. It can be difficult, even for the most dedicated environmentalist, to determine what works and what doesn't and where to prioritize expenditures. Costs and results provide feedback that encourage people to change strategy when necessary. If you don't feel the direct cost of your decisions, the feedback you receive is secondhand and less likely to be effective.

Using incentives to promote environmentally sustainable choices isn't new. Technology makes those incentives clearer and helps respond to them more effectively. Before I show how, let's look at a low-tech example of environmental incentives that is already saving energy and water.

ENVIRONMENTAL INCENTIVES MAKE SUSTAINABILITY MORE SUCCESSFUL

I'm not bragging, but I probably have the world's largest collection of hotel "green" cards—the ones you find in your room telling you how to help save the planet by reusing your towels or linens. Some people steal hotel bathrobes. I steal the little cards (it's not kleptomania if it's for a good cause) that include guilt-inducing phrases like "In addition to decreasing water and energy consumption, you help us reduce the amount of detergent waste water that must be recycled within our community." The cards have a variety of ways to tell you that unless you reuse your towels you hate the planet.

One says, "We are committed to energy and resource conservation. Every day, tons of detergent and millions of gallons of water are used to wash towels which have only been used once." The card explains you can help reduce that impact: "A towel that is hung up means 'I'll use it again.'" They say they are committed, but you are the one doing the work.

Another card says, "If you would like to help save our resources, please leave this card on the bed in the morning and we will make your bed without changing your linens." When they say "our resources," it is unclear whether "resources" means water and energy or their money. Either way, the money (and resources) being saved goes to the hotel, not to you. Some hotels make reusing your linens the default, instead of an option. Instead of opting out of service, they require you to opt in. "Your Sweet Dreams® bed linen will be changed every third day. If you would like your linen changed sooner, just place this card on your pillow." This probably works better—saving more water and money for the hotel—but it still relies on guilt and doesn't share the benefits of those savings with the guest.

The problem with each of these is that while guilt works, providing a small incentive to conserve significantly improves the participation in conservation programs. Hotels save money by using fewer resources and reducing the labor necessary to clean rooms, so why not share a bit of that benefit? That is exactly what Westin Hotels thought, so they decided to provide a small reward to guests who forgo having their room made up. Westin's program uses a bit of carrot and a bit of stick. Much like the hotels that use the guilt-centered approach, the card informs guests they will save, "37.2 gallons of water, 25,000 BTUs of natural gas, 0.19 kWh [kilowatt hours] of electricity, and 7 oz. of cleaning-product chemicals." But they add a financial incentive as well. "You'll receive $5.00 to spend at participating food and beverage venues or 500 SPG® Starpoints®." That simple change made a dramatic difference.

The people most responsive to the incentives are business travelers who want privacy and are less concerned about making the bed. They also wouldn't mind a few extra points from their favorite loyalty program. As the *New York Times* noted, "While leisure guests occasionally opted in, frequent business travelers 'ate it up,' said James Gancos, the chief executive and founder of the Guestbook, a loyalty program for independent boutique hotels, and the former hotel manager at the Sheraton Seattle."[52] Business travelers told Guestbook, "I love being green, I love the extra points, I didn't want people in my room anyway."

The program also appeals to younger travelers. Speaking to the *Times*, Adam Weissenberg, the global leader of travel and hospitality for Deloitte, said hotels that were changing linens every day were receiving negative feedback. "They received criticism from younger travelers. 'This is ridiculous that they're changing my towels and sheets every day. I don't need that; it does harm the environment.'" In addition to saving money and resources, the hotels had an incentive to provide options to accommodate their customers. Thirty-eight percent of travelers told MMGY Global's Portrait of American Travelers 2017–18 that they "would be willing to pay more for a travel service provider who demonstrates environmental responsibility, a 13 percent increase from 2014." For guests looking for environmentally friendly options, Delta Hotels in Canada provided an alternative to the loyalty points. Stefan Lorch, vice president of Guest Experience at Marriott, Delta's parent company, told me they offered a program called GreenSTAY that plants one tree for every night guests opt out of housekeeping. "They have quite a few guests who don't take the points, but take trees," he noted. Delta celebrated planting its 100,000th tree in 2017.

Providing incentives resulted in meaningful savings for Marriott and other participating hotels. Marriott says it reduced energy use by 13.2 percent, water use by nearly 8 percent, and green-house gas emissions by nearly 16 percent. But that is not the only reason hotels are using incentives to change guest behavior. "In California, we have a drought situation," Lorch told me. "Any gallon of water we can avoid, that is a win for the entire community. We are seeing this as a necessity and a requirement."

Others are noticing the success. IHG, which includes InterContinental Hotels and Holiday Inn, offered something similar. Their card encouraged people to participate, noting that "caring for the environment has never been so easy" by offering 500 IHG reward points for a stay of two or more nights. In Portland, Oregon, the Provenance hotel offered three levels. "Old-School" is the opt-in system where you request new linens by putting the card on the bed. To reach the "Eco-Friendly" level, you agree that linens won't be changed and any towels that are hung up won't be laundered. The top level, "Tree Hugger," declines all housekeeping. They note, "We reward Tree Huggers. Each day you opt out of housekeeping service by placing this on your guest room door you'll get $5 to spend in the honor bar." Suddenly that $6 Snickers bar seems more reasonable. The Essex Resort and Spa in Vermont, where my wife stayed and made a point to bring back the card for my collection (I admit I have a problem), also offers five dollars a day to decline housekeeping services.

Each of these hotels learned what economists preach every day—incentives matter. Provide even a small incentive to save resources and people will adjust their behavior. Rather than using guilt to push people into behavior that saved money for the hotel—but not for guests—sharing the benefits of conservation engaged guests in a simple but powerful way. Although these programs are straightforward and do not need smartphones to implement them, they highlight two ways smartphones and personal technology can expand these opportunities.

First, rather than a top-down approach, the hotels use individual incentives to engage their guests. Hotels can't know the personal preferences of each of their guests. With incentives they don't need to. If a guest doesn't mind using the same towels and linens for a few days, they can make that decision themselves. Without an incentive, however, that same guest may decide to have their room cleaned every day because it costs them nothing and all else being equal, clean sheets are nice. Provide a small incentive and suddenly that same guest decides the loyalty points are more important than a crisply made bed. The only person who can make that calculus is the guest. There is no one-size-fits-all.

Second, these programs take many small decisions and aggregate them into meaningful results. Each 37.2 gallons saved by a guest adds up to an 8 percent reduction in water use. Each Delta Hotels guest who skips housekeeping plants a tree, which soon results in 100,000 more trees across Canada. Incentives have always made a difference in how people behave. In a smartphone world, existing incentives can be made even more powerful. Rather than simply relying on guilt, smartphones help make changes that are durable and reward people for using resources more efficiently.

Some understandably fear that such incentives can be used by politicians to "nudge" people in a particular direction, punishing politically incorrect behavior and subsidizing behavior that is pleasing to politicians and planners.[53] Alternatively, combining incentives and information can empower individuals to make their own decisions, rather than giving authorities more control. Giving people incentives and choices moves power out of the hands of regulators and politicians—who don't feel the cost of their decisions—and into the hands of individuals who feel the costs personally and can make decisions based on real-world circumstances. The best environmental choices are made closest to the source, and using smartphones to offer incentives can increase personal freedom, accountability, and environmental responsibility.

There is interesting research that demonstrates how powerful incentives can be when combined with information technology.

INCENTIVES ARE MORE POWERFUL THAN MORAL SUASION

Open your gas or electric bill and you may find the utility has compared your energy use to that of your neighbors. A little competition, it turns out, can influence the amount of energy people use. A company called Opower sends customized energy reports that "feature personalized energy use feedback, social comparisons, and energy conservation information" in an effort to encourage homeowners to conserve. Researchers found consistent reminders decreased energy use of high-end consumers by about 6 percent.[54] These approaches can work, but they are limited. The study found it was most effective on the highest-consuming energy users. Researchers also found that without reminders some people exhibited backsliding, losing some of the gains that had been made.

Electricity conservation is important for both economic and environmental reasons. Utilities must ensure they can meet the peak demand during the day, usually in the late afternoon or early evening. To do that, utilities must either build new power plants or purchase energy on the spot market. Following the laws of supply and demand, when demand is highest, so too are prices. If a utility can avoid buying electricity when it is most expensive, they can keep electricity rates low for everyone. Additionally, peak demand must be served by generation that can be turned on quickly. In many parts of the world, the most dispatchable form of electricity is generated by coal or natural gas. Reducing peak demand and the need for fossil fuel–based dispatchable electricity can be a win-win for the environment and consumer pocketbooks.

Currently, consumers have only limited incentives to reduce energy use during these critical peak hours. In many places, electricity rates are higher during the day but do not reflect the true costs of using electricity during hours when prices are highest. The head of the smart-meter program at the US Department of Energy told me that in 2018, only about 2 percent of ratepayers saw time-of-use rates.[55] There is a reason electricity customers have been protected from spikes in energy prices. Consumers need to receive the information in a timely way and then figure out how to respond. For more than a hundred years, this type of information simply hasn't been available, so regulators protected consumers from price swings, putting in place electricity rates that had a mild and predictable rate structure, spreading the costs of peak demand more evenly across the day. Consumers pay less than the real price during peak hours and pay more during other parts of the day to make up the difference. If consumers realized how costly that peak-hour electricity actually is they could find ways to conserve or move demand to off-peak hours.

Rather than relying on guilt to shift demand away from expensive peak hours, incentives delivered through smartphones are more effective at promoting energy conservation during those critical parts of the day.

Three researchers in Japan set up a system where nearly seven hundred households were chosen to participate in an experiment comparing the effect of moral suasion (i.e., guilt) and incentives.[56] Everyone participating received an advanced electricity meter and an in-home display. Explaining the need for their research, Koichiro Ito, Takanori Ida, and Makoto Tanaka noted, "Electricity consumers generally do not pay prices that reflect the relatively high marginal costs of electricity during peak-demand hours. This mismatch is a fundamental economic inefficiency in electricity markets." Participants were broken into three groups. The "moral suasion group" received the meter and display and occasionally received messages encouraging them to conserve. The "economic incentive group" received the meter and display and messages offering a monetary incentive to conserve. The final group was the control group and received the meter and display but no messages.

Ito, Ida, and Tanaka tested two ways to encourage consumers to reduce electricity use during peak hours. For the moral suasion group, they sent a message saying that conservation "will be required for the society in 'critical peak-demand hours.'" The group receiving the economic incentive received a message reading, "Notice of Demand Response: In the following critical peak-demand hours, you will be charged a very high electricity price, so please reduce your electricity usage: one to four pm on Tuesday, August 21. The price will be 85 yen (+ 60 yen) per kWh." The economic incentive group received a price increase of 40, 60, or 80 cents per kilowatt hour (kWh), a significant increase over the existing base price of about 25 cents per kWh.

The results were very clear. The researchers found that "the level of reduction is much larger for the economic incentive treatment." The incentives were so effective, they even worked outside the period of peak demand. "These results imply that the economic incentives in our experiment motivated customers to lower their usage in both the non-treatment hours and the treatment hours," they reported. Once consumers felt the cost of their energy use, they found ways to conserve and made it a habit. They applied the new information they had about their own personal energy use to reduce demand in a way that uniquely suited them.

By way of contrast, the moral suasion approach had a limited effect. The study found the "moral suasion effect was statistically significant in the first cycle but became insignificant in the remaining cycles for the summer." People simply became inured to the guilt and reverted to their old behavior. As with the program comparing you to your neighbors, or efforts by the hotel to get you to reuse your towels, guilt only goes so far.

Smartphones provide immediate information and play an important role in making these incentives tangible. One of the researchers examined similar economic incentives in California, but the information was available to consumers only on their monthly electric bills. The effect was much smaller. In comparing the two methods—immediate information delivered by text message compared to monthly, aggregated information—they noted, "Customers in our experiment had access to salient information about their real-time marginal price via in-home displays and text messages, whereas Californian customers in Ito's study received their price information only through their monthly bills. Although monthly electricity bills provide information on marginal price, such information is not usually transparent, and consumers receive it with a month lag." As a result, California consumers didn't make meaningful changes. The combination of an economic incentive and immediate information made the difference. Timely information serves as a reminder to make small changes, like waiting to dry your clothes, and the encouragement to take advantage of opportunities to save and pay attention to conservation amidst the competition for our attention.

THE POWER OF HAVING SKIN IN THE GAME

Conserving electricity, and reducing the environmental impact associated with electrical generation, offers some of the low-hanging fruit for smartphone-connected small environmental technologies. Every kilowatt hour of electricity has a price, and that cost encourages conservation. Making information more available takes existing incentives and makes them more powerful. The traditional approach to cutting pollution is to increase the price until the impact declines, as we saw with sulfur dioxide when the federal government put a price on it to reduce acid rain. Less appreciated, in part because it has been difficult, is the ability of technology to make existing prices more obvious and meaningful even if they aren't increased. You may not know the train is coming until it blows the whistle. Smartphone technologies can be the train whistle that draws your attention to existing costs you might not otherwise have noticed. The power of smartphone-connected technologies is that they magnify existing incentives and align them with the ability to act. This combination is a more effective way to guide us to good environmental outcomes.

Nassim Taleb, a former option trader and risk analyst, argues that having a personal stake in the outcome is critical to ensuring the best policies are adopted. He writes, "There is no possible risk management method that can replace skin in the game in cases where informational opacity is compounded by information asymmetry…that arises when those who gain the upside resulting from actions performed under some degree of uncertainty are not the same as those who incur the downside of those same acts."[57] The difference skin in the game makes is evident in the contrast between Sense and the "smart meters" that have been deployed by utilities and the federal government over the last decade. In 2009, President Obama promised that smart electricity meters would be installed in homes across the country, allowing them to "actually monitor how much energy your family is using by the month, by the week, by the day, or even by the hour."[58] Ten years later many utilities had not yet installed the meters and by then the technology was dated. Smart meters were intended to save money by replacing meter readers with remote monitoring. They can also report on electricity use every fifteen minutes.[59]

By way of contrast, Sense detects your electricity use millions of times a second and provides the detailed information necessary to determine where homeowners can conserve. In the fifteen minutes between the times a government-sponsored smart meter reports data, Sense measures electricity use nearly one billion times. The difference between smart meters and Sense is that smart meters are primarily designed to help utilities calculate electricity bills and prioritize the needs of utilities rather than customers. Sense, on the other hand, is used by homeowners who pay the bills and are accountable for the costs of their decisions. Who pays the price or receives the benefit makes the difference in the functionality of the technology.

The direct feedback and accountability provided by technologies like Sense aren't always present when "experts" offer policy advice, leading bad policies to endure and ideological biases to override real-world results.

Philip Tetlock studies what makes people good at predicting future events and identifies the pitfalls that lead to poor predictions. He warns that a lack of accountability means that bad predictions get repeated. "The consumers of forecasting—governments, business and the public—don't demand evidence of accuracy" or hold people accountable for that lack of accuracy. "So, there is no measurement. Which means no revision. And without revision, there can be no improvement."[60] Incentives are critical to improving accuracy over time. We measure what we value, and incentives help ensure we value conservation and energy efficiency.

Rather than try to predict the future based on an incomplete understanding of the complexity and without accountability for failure, we should look to environmental technologies that are flexible and give individuals a stake in the outcome. This is the lesson of Ito, Ida, and Tanaka's experiment in Japan. The Sense monitor is just the tip of the iceberg. New technologies create opportunity for larger, systemic changes.

Steve Barrager, whose first job out of college was working on the Apollo space program, cowrote a book on the future of electricity, and he is excited about the changes that are coming and the ability to build new incentives into electricity. Called "transactive energy," it would allow people to buy electricity in the same way they buy smartphone plans, choosing the amount and rate structure that fits them. "What you want to do, if you are going to make the system more efficient, you are trying to optimize net benefit to customers," he told me.[61] "In a market economy that comes from prices. Customers decide what they are going to do in a way that maximizes their net benefit." Everyone is different, but the current system treats everyone virtually the same. Electricity customers fear that, without useful information, changing the rate structure would simply end up costing them more. One study found, "Only a quarter of consumers today feel that their utility is connecting the dots for them—that the utility is properly informing them about how much energy and water they use, and then correlating that use to both its environmental and economic impact on the consumer."[62] With new technology available, people are coming to expect this type of information.

In a survey of the rapid pace of disruptive innovation in 2015, McKinsey and Company highlighted several trends that are putting more information in the hands of consumers. As a result, "the former passivity of customers has been superseded by a desire to fulfill their own talents and express their own ideas, feelings, and thoughts."[63] That trend has become stronger since then. That is exactly what Mike Phillips of Sense says they are trying to do. "I think of it as consumer-facing applications and tech." In the past, consumers were resigned to the reality that others would make decisions for them, simply because the information was not available. Now that the information is available, it opens up a huge range of options.

It isn't only the consumer-facing companies who see these changes and the opportunity. Utilities are also beginning to think about how this could achieve some of their goals, while keeping consumers happy. In 2015 utility experts warned, "Driven primarily by innovation in technology and changing societal values, choice has extended to areas that in the past had few or none. Think smartphones, social media or the rise of the new sharing economy. The Internet has made customer choice as simple and as powerful as the click."[64] The transformation that has occurred since that warning is remarkable.

First, it will mean more transparency. Ask someone right now how much they pay for electricity and the chances are they won't know. To make it even more complicated, there are different rates during the daytime and at night, and those rates change after you pass a threshold for the month. Making transactions and prices transparent will be key to changing the way our electrical system works. Steve Barrager noted, "There are 17 different ways of calculating prices and you have to check the tariff schedule. It is virtually impossible to figure out. With transactive energy you can see every flow and transaction." That transparency allows consumers to know exactly what they are using and what they are paying. Barrager's coauthor Edward Cazalet says this is critical if we want to give people more options. "Move to a system where you subscribe to power to satisfy most of your needs. Subscribe based on previous meter needs. Contract for an amount of power that roughly fits your current needs. Customers like subscriptions, fixed monthly payment."[65]

Second, it will create more fairness. The ideal rate structure for electricity allows everyone to access the power they need at prices they can afford. Low-income families can choose inexpensive options. People who don't need electricity during peak hours—or can cut back—have incentives to do that. As Barrager notes, "The current rate making system is very bureaucratic and very politicized. We really don't know if we are distributing resources fairly." Each consumer is different and only they know what suits their needs and budget. The complex pricing systems created by utility commissions attempt to cater to different groups, but fundamentally, they are the same as they were fifty years ago. In many places, electricity prices still haven't caught up with the opportunities that smartphones are providing.

AI TO SAVE ELECTRICITY

Putting information in the hands of those with skin in the game is a powerful first step, but it may not be enough. Even those motivated to monitor their energy use can only do so much. Time also has a value and spending too much of it to save a few kilowatt hours doesn't make sense for very many people. All that information can be overwhelming. If users feel they need to constantly change their thermostat in response to electricity prices, they may decide it just isn't worth it.[66] Artificial intelligence is helping fill that gap.

Microsoft launched "AI for Earth" precisely to find ways to use artificial intelligence to address environmental problems. Microsoft's goal is to offer their expertise to organizations working on difficult environmental issues. Those organizations may understand the problem but don't necessarily know how to apply artificial intelligence or have the skill to make that happen. For that reason, the primary audience for Microsoft's project isn't large organizations but rather small- and medium-sized groups that can apply AI with personal technology to a specific problem.

Artificial intelligence is helping consumers in several ways. The growing number of appliances connected to the internet is called the Internet of Things, where connectivity is being extended even to lightbulbs. With that connectivity comes a flood of data— data that can't meaningfully be interpreted and managed by individuals. Guided by simple preferences chosen by users, AI is turning that flood of data into useful decisions. Addressing the opportunities for businesses to reduce waste and increase profits, business writer Adam Uzialko reported, "Since IoT can be used at the circuit level to collect data about a system performance in more detail than traditional monitoring systems, much more data about its operation is produced and therefore AI is also necessary to analyze and interpret said data in a timely manner. Artificial intelligence can also automate certain functions to save energy without extra efforts, like making adjustments to climate control and lighting to save energy."[67] AI can adjust your thermostat and even control appliances, like dryers, to reduce electricity costs.

Although we tend to think about artificial intelligence in terms that conjure up big computers sorting out extremely complicated problems, it may be most useful completing simple tasks more effectively. Rather than doing one thing 100 percent better, it does 100 things 1 percent better. "AI makes the difference even for simple tasks," Rob Bernard of Microsoft's AI for Earth told me. "People didn't program their VCRs. Now, people watch on-demand all the time. The ideal state is that even people who do nothing are using fewer resources." AI makes small improvements possible because they reduce the cost of action. Improving efficiency by 1 percent can be difficult because it requires attention and effort that would normally be more than such a small improvement is worth. With artificial intelligence, the cost of these small improvements is so inconsequential, they suddenly make sense, and a huge range of opportunities is opened.

The best examples of this new opportunity are smart thermostats. Thermostats that automatically adjust home temperature at certain times of the day are not new. As technology improved during the late twentieth century, utilities and the federal government promoted thermostats designed to reduce electricity demand. Programmable thermostats allowed homeowners to set different temperatures for when they were home or away. The goal was to reduce energy use when people weren't in the house. It seems like a fairly straightforward task. It wasn't.

The Environmental Protection Agency (EPA) rescinded the Energy STAR ratings for all programmable thermostats in 2009.[68] In their announcement, EPA noted, "Significant questions have been raised as to the net energy savings and environmental benefits being achieved with the current set of ENERGY STAR qualifying" thermostats. One utility testified that "programmed thermostats resulted in 25 percent higher summer peak demand compared to homes with non-programmed thermostats."[69] Programmable thermostats failed to save energy for a variety of reasons. Because "many people end up switching their system on and off manually, programmable thermostats might cause most people to use more energy," explained Kamin Whitehouse, a computer science professor at the University of Virginia.[70] The failure of programmable thermostats created the opportunity for something else to replace it.

What emerged was learning thermostats like Nest and Ecobee. Developed by Apple programmers and later purchased by Google, Nest combines the ability to control the temperature using a smartphone with artificial intelligence that learns how you live, calculating how to help you stay comfortable and save money. One of Nest's creators, Tony Fadell, was building a new home and wanted to control the energy systems in the house using his phone. When he found that such systems were extremely expensive, he decided to change that. The thermostat must be designed to "be smart enough to learn his routine and to program its own schedule accordingly, or to switch off automatically if he went out." Although people weren't very good at programming their thermostat, they were used to engaging with them. Patty Durand, president of the Smart Grid Consumer Collaborative, summarized, "Smart thermostats are probably the best gateway to engage consumers. If you're going to pick one thing, the research points towards thermostats."[71]

As people adjust the thermostat, Nest learns their habits, figuring out when they want the room to be comfortable and when they are out of the house, reducing energy use during those times. The Nest can also track the location of your phone to determine when you are out of the house. Both Nest and Ecobee use sensors to determine whether people are at home and can send notifications to users' smartphones asking them if they would like to adjust the setting.

Finally, the artificial intelligence in Nest can adjust warming and cooling to take advantage of the weather forecast. For well-insulated homes that hold temperature effectively, Nest will precondition a house when it is easiest, rather than waiting until a home is too hot or cold and then using large amounts of energy to catch up. This simple step can make a big difference. Marissa Hummon is a Harvard-trained physicist who worked at the National Renewable Energy Laboratory and designed technologies that make it easier for customers to access renewables and conserve electricity. She has been at the cutting edge of these innovations for nearly a decade. Technology, she argues, is well suited to make this kind of simple but meaningful change. "Cooling the house down early in the morning at 80 degrees rather than later at 100 degrees," she told me, "saves a ton of energy. Thermal mass can maintain a 3-degree range over six to eight hours. If you cool the house early, then it stays that way most of the day." This is exactly what smart thermostats like Nest and Ecobee can do—plan the energy use for the day so it keeps a building comfortable efficiently. "If you can keep the temperature between 73 and 76, while avoiding peak load in the afternoon,"

Hummon notes, "then the consumer is happy, and the utility is happy because they didn't have to meet peak load."[72]

Ecobee and Google Nest believe the combination of sensors and AI generates significant energy savings. Ecobee estimates its users had saved 30 terawatt hours of energy by the beginning of 2018. Ecobee users, they say, are saving up to 23 percent on their electric bills. Google Nest estimates that its users saved more than 76 billion kilowatt hours between October 2011 and April 2021—more than a month's worth of coal-generated power in the United States.[73] Although these claims are not independently verified, several studies demonstrate that smart thermostats are yielding big savings.

Utilities, which have a mandate to keep costs low and encourage conservation, are turning to smart thermostats to deliver what smart meters have not. In Oregon, Portland General Electric (PGE) compared the effects of providing smart meters using Advanced Metering Infrastructure (AMI) and Nest smart thermostats. PGE gave some of its customers AMI meters and incentives to cut electricity use by increasing prices during peak demand, known as "Time of Use," or TOU, pricing. The meters provided information about electricity use at regular intervals. The meters did not, however, learn or use AI to respond to those prices. PGE sent prompts to users and found that "more effective communication" was essential to realizing the full benefits they hoped smart meters would provide. After their initial test, PGE reported that the TOU rates were "complex and difficult to understand" and resulted in very little savings.[74]

PGE then launched a pilot project to see if Nest thermostats would be more effective. On a voluntary basis, customers signed up for a program called "Rush Hour Rewards" which provided rebates for allowing the utility to adjust their thermostat for three hours during periods of high demand. The utility warned customers in advance that their thermostat would be adjusted, allowing them to override the change. If they did nothing, the Nest thermostat would precondition the house when electricity demand was low and then automatically adjust the thermostat when demand and prices increased. That preconditioning is a critical element of AI's value. Homeowners can't take the time to think two hours in advance to set their house to an appropriate temperature, anticipating that it will be more expensive later. We don't think about the temperature of our house until we are uncomfortable. Artificial intelligence can.

PGE compared the energy savings of those who allowed their thermostats to be adjusted to a control group that did not receive the Rush Hour Rewards signals. Although homes with smart thermostats had similar use patterns outside the events, those in the Nest pilot saw significant reductions during peak demand. During the winter, combining incentives with artificial intelligence reduced average demand by 20–33 percent during the first event hour, 15–30 percent during the second hour, and about 10–25 percent during the third hour. During summer, the pilot reduced demand by about 40 percent during the first hour, 30 percent during the second hour, and 20 percent during the third hour.[75] Customer surveys of participants reported high satisfaction, and it "received high average ratings of 8 or greater on a 10-point scale from treatment and control group customers." When the Public Utility Commission of Oregon subsequently evaluated PGE's projects, the commission highlighted the success of the smart thermostats, encouraging PGE to "more aggressively augment the Direct Load Control Thermostat (DCLT) program offering, including exploring system wide direct installation of smart thermostats."[76] They also encouraged PGE to expand the program to "include customers with other brands of connected thermostats." Comparing the results from smart meters and smart thermostats, PGE and the Oregon PUC found the smart thermostats yielded better results. Other jurisdictions had a similar experience.

As part of a larger TOU rate study, Southern California Edison (SCE) tested the effectiveness of Nest thermostats at reducing peak demand. During the two years of the study, users were provided with TOU price signals. During the second year, Nest users also used a program called "Time of Savings."[77] Unlike the Rush Hour Rewards program, which provides a rebate for lower use, SCE used the Time of Savings program, where the thermostat calibrates energy use to the electricity rates provided by the utility. Households with smart thermostats performed better than households receiving only the price signals. SCE reported, "Households who had previously purchased smart thermostats reduced summer 2017 peak period usage by approximately 6.7 percent, which was significantly higher compared to...load reductions of 3.7 percent" for those without smart thermostats.[78] The following year, Nest also used rate data in the Time of Savings program, which "significantly increased the magnitude of peak load reductions relative to the first summer." The researchers point out that due to weather variability and other factors, comparing one year to the next is difficult. Even without the additional Time of Savings effect, however, SCE's pilot saw positive results from Nest's ability to automatically respond to peak pricing.

Artificial intelligence isn't just about reminding people. It can plan ahead with a sophistication that is difficult for busy people. It can rapidly incorporate price information and respond with a timeliness that is beyond virtually everyone who isn't staring at market electric rates in real time (if you do feel like staring at them, however, there are apps for that too!). Incentives work. Incentives plus an assist from AI work better.

"Compared to the 'smart meter' the Nest thermostat is genius," wrote Barrager and Cazalet in their book on the future of energy. "After the homeowner has adjusted it ten or fifteen times it begins to understand what it is supposed to do. It begins to anticipate when the building should be warmed and when it should be cooled." Really, though, what Nest does isn't that complicated. Computationally, it takes data and looks for patterns and adjusts to signals—whether that is a price signal, adjustments from a user, or a signal from a phone that you have left your house. The signals could even be environmental, notes Marissa Hummon, and your system could respond to "a greenness signal as opposed to a monetary signal." One utility in the United Kingdom and Texas that we will learn about later is doing just that.

The system is being refined all the time. Making the electrical system more responsive and providing more options about the type of energy we want and the price we pay is now possible in a way it wasn't just two decades ago. In 2014, Barrager and Cazalet wrote, "Today, it is possible for a single customer to have the same communication and information technology power that the [utilities] have. Customers have access to decision-making processes that are as sophisticated as producers and [utilities]." We've moved far beyond that. What's more, that progress is occurring outside the political system. Customers are buying smart thermostats because they save money. Utilities are engaging with smart thermostats because they help keep costs low and customers happy, and they reduce the need for risky new expenditures. The sophistication of smart thermostats is improving constantly. The upside is still enormous for customers and utilities. Smart thermostats demonstrate how incentives and artificial intelligence are beginning to create opportunities to make a dramatic leap in the way we use electricity.

AI TO SAVE WATER

For some people, the desire to save electricity, money, and the environment is what spurs their innovation and use of AI. For Keri Waters, it was the love of a hot shower.

California had been dealing with drought for several years. In 2015 it was particularly bad, even by California standards. California's water problems go back decades and in some places more than a century. The book *Cadillac Desert*, which was called "the definitive work on the West's water crisis," was published in 1986.[79] A lot has happened since 1986, and fights over water have become more intense. It is in that context that Keri Waters decided to act.

65

Already a successful Silicon Valley entrepreneur, Waters wondered how she could help address the problems facing the state. One key data point stuck out to her. "In the middle of California's record-breaking drought in 2015, as the state asked residents to cut water use 20%," reported the tech blog Fast Company, "Santa Cruz–based entrepreneur Keri Waters noticed another statistic: in a typical American house, around 10 percent of water used is lost to leaks."[80] The ability to detect water leaks in your house, however, was almost nonexistent. "I read an article about 10 percent of household water is lost to leaks," Waters told me. "My first reaction was righteous indignation. Why am I taking shorter showers—this is pleasure in my day—when I am wasting 10 percent? I have no idea how much change the other things I'm doing are making." She wasn't the only one interested in reducing the impact of leaks.

Water districts across the country began looking for ways to help customers identify and stop leaks. Sammamish, Washington, is one of the most technologically connected communities in the country. A wealthy suburb of Seattle, it is commuting distance from Microsoft's headquarters as well as Amazon's and Google's offices in the Seattle area. The Sammamish Water District launched a program to give everyone access to daily water usage information and the ability to set usage or price targets using an online web portal. When you are nearing those targets, the system will send an e-mail or push notification to let you know. One key element is the ability to notify consumers that they have a leak. In a presentation to the water district board in February of 2018, the staff singled out one unlucky colleague who had a running toilet that cost eight dollars for just one day's waste. Homeowners who don't hear their toilet running, or aren't home, may find they've wasted hundreds of gallons of water, only learning about it when they get an exorbitant bill at the end of the month.

The water district saw several advantages to the online portal. First, it was preferable to the district calling people when they detected a leak. Water district commissioners expressed concern that the calls had a "Big Brother" feel to them. An email notification doesn't feel the same way. Second, it helps the district conserve water and keep rates low. Rather than people getting sticker shock at the end of the month, leaks or running toilets can be addressed relatively quickly. In a community like Sammamish, where people are accustomed to using technology, an online portal could be a good way to connect with the water district's customers. Finally, there is the simple conservation ethic. Although it is known for rain, the Seattle area still faces conflicts over water use, and what isn't used by people can be left in streams for salmon, because that additional water keeps streams cool.

The system is far from perfect. The water district began developing the system in 2017. By 2019, it was still not deployed. There are, of course, private-sector companies who experience delays. The water district, however, is not going to lose customers if the portal isn't completed on time. Indeed, because so few water districts use this kind of technology, the district deserves credit for the experiment. Nonetheless, the portal's launch was repeatedly pushed back. Further, although the system will notify you if there is a serious leak, there isn't anything you can do about it immediately, unless you go home or have someone fix the problem. The information is helpful, but it doesn't offer the ability to act.

By way of contrast, Keri Waters decided to create a solution called Buoy, a small device users attach to their home's water line and connects to a smartphone. Not only did it track water usage; it also allowed users to shut off the water to the house with a click of a button on their phone in an emergency. Artificial intelligence is an important part of its functionality. It was so important the company's first web page address was buoy.ai.

Like Sense or Nest, the Buoy provided users with information about how they use water and gave them increased control over that usage. The artificial intelligence analyzed the flow of water and determined whether it was going to your sprinkler, a washing machine, or your morning shower. By tracking where the water went, Buoy provided information that could reduce water waste and, potentially, water shortages that worried Buoy's creators. "The software is similar to Sense in terms of disaggregating use in your home," explained Waters. "We know what time and what flow rate, and we can use the signatures to determine where you are using that water. Our accuracy is higher than those who do this for energy. We use water in very similar ways. We all have showers, sinks, dishwashers and washing machines." The algorithm can also detect small leaks that can eat up a significant amount of a household's water use. Eliminating those leaks can add up. When the company began testing, they found users wasted about 9.5 percent of water. Fixing those leaks would save up to 13,000 gallons of water a year per household, saving hundreds of dollars. As water becomes more expensive, eliminating waste becomes more valuable.

Losing 10 percent of water a year to leaks is significant, and the little leaks are difficult to catch without some help. Waters offers of a list of ways people waste water. "Broken sprinkler heads. People are astonished by leaking toilets that leak hundreds of gallons a day. Appliance hoses popping off the back." Someone might walk past these small leaks and think nothing of them, assuming they don't really make a difference. Buoy makes it clear that every drop of water matters. "When you see the data, you get around to fixing it."

Buoy users could also shut off major leaks immediately by pressing a button on their smartphone. "The insurance industry loses $10 billion a year from leaks," noted Waters. "About 2 percent of houses every year will have a catastrophic leak that will lead to an insurance claim, typically costing about $10,000." The Buoy was able to detect a major leak and notify a user on their smartphone in under a minute. With a click of a button on the Buoy app, a homeowner could shut off the water to their house before the leak did major damage. Rather than coming home to a flooded basement, users tell the device to shut off the water.

The benefits of the technology aren't limited to the user, however. Water districts and insurance companies can also benefit.

Former Santa Cruz Mayor Hilary Bryant joined Buoy, drawing on her experience with the city's water district and planning for the combination of growth and water scarcity. Even with improved metering technology, she said that most water districts "are still in a position where, if there is a leak, they send a letter." The letter arrives several days after the leak began, and only offers suggestions of where to look for the leak. Customers are left in the position of being told they have a leak—and have had one for days—but they don't know where. Buoy, on the other hand, could pinpoint the cause. "Buoy customer service becomes their customer service," noted Bryant. Not surprisingly, Waters and Bryant argue water districts would benefit by offering rebates to buy a Buoy, in the same way electric utilities offer rebates for buying a smart thermostat. Of course, their recommendation is self-serving, but that's not a bad thing. The company has a financial incentive to make sure their users are happy and save money. The better they do at reducing leaks, the fewer frustrated calls water districts receive, and the more people will buy their product. Rather than each district trying to develop systems on their own, they can piggyback on the technology developed and refined by Buoy.

As water becomes more expensive, homeowners will need more information to reduce waste and conserve water. There are only so many low-flow showerheads and toilets people can install, and even the most frugal water user can waste all that effort with an undetected leak.

Buoy was one of the first in a wave of tools designed to help homeowners reduce water waste. Another system similar to Buoy, called Phyn, tracks water use 240 times a second so it can determine what appliances are being used and identify the source of leaks.[81] Another device called Flume attaches to a water meter and reads the magnetic field from the spinning sensor to calculate water use, and then applies artificial intelligence to determine how water is being used and detect leaks.[82] WaterSmart and Dropcountr both work with utilities to provide information to users about their daily use. That information will be necessary to help users find better ways to save water, reduce their water bills, limit the impact of drought, and enjoy their morning shower without guilt.

Interestingly, like many leading-edge innovators, Buoy didn't survive the growing competition and was discontinued in 2021.[83] Entrepreneurs constantly face the forces of "creative destruction" that are part of markets, with innovations that break new ground falling prey to the next improvement. That progress is good for the environment, as each new iteration makes it easier to save water and other resources. "I think that water saving technologies are going to continue to become more and more mainstream," Keri Waters told me in 2021, "because we are seeing water challenges in the West and that will not abate."[84] So, while the Buoy is no longer available, Waters's inspiration and efforts continue to have a positive impact on the environment.

AI IN THE FIELDS

Drought and water restrictions are inconvenient for homeowners whose lawns are brown and who may have to cut back on long showers. For farmers they can be devastating. Using every bit of water wisely, from either irrigation or the sky, is critical to help farmers compete in international markets and deal with commodity prices over which they have no control. This is one reason AI for Earth has focused on helping farmers. "We'll help farmers put AI to work not only to better analyze soil and rainfall conditions, but also to use predictive analytics to improve agricultural yields and reduce adverse environmental impacts," said Microsoft president Brad Smith when launching AI for Earth.

Long before Microsoft launched AI for Earth, two researchers at Washington State University were already demonstrating the ability to use simple data points and apply artificial intelligence to help farmers use water effectively. Irrigation researcher Troy Peters and programmer Li Tan teamed up to create the "Irrigation Scheduler Mobile" app, which helps farmers plan for rain and schedule how they use water.[85] Farmers enter some basic information, including the location of the farm and the crop. The app does much of the rest. "It has the National Geological survey and has the soil composition at that location," Li Tan explained. The model allows it to determine how much water the soil can hold. The app also connects with local weather networks to plan for rain. Irrigation Scheduler aggregates this information to develop a watering plan for farmers that applies water for the particular crop. Water is valuable and many farmers can intuitively calculate water use, adjusting for many variables. This is nothing new. Operation of the US National Weather Service was transferred from the military to the US Department of Agriculture in 1890 to provide farmers with accurate forecasts. AI can help farmers do many things that 1 percent better. Even if rapid access to weather data combined with artificial intelligence offers only a marginal improvement over the wisdom of an experienced farmer, those margins matter in the commodity market.

Even without artificial intelligence, personal technologies help farmers plan how they use resources. Kyle Bridgeforth is a fifth-generation farmer in northern Alabama, where the farm he manages with his brother and cousin grows soybeans, cotton, corn, and wheat. Unlike past generations, he is barraged with data about his crops. That has fundamentally changed how he operates. "It is great, especially for someone like me," he says. "Over the last generation, farming has made that transition from where it was an art—where you had to know the land and your equipment worked by trial and error—now it has become a science." Smartphones put that information at their fingertips in the field. He notes that even the basic apps, such as the weather app on an iPhone, can be extremely useful. "The native app for weather includes wind speed. Every operator uses it all day long to make sure chemicals aren't spraying where they aren't supposed to." There's an app that includes the safety labels for the various chemicals they use. These seem simple, but Bridgeforth notes that without the weather app he had to guess the wind speed based on experience.

The data he uses go far beyond that. Sensors allow him to adjust how much seed is being used based on soil samples every twenty feet. They tell him the moisture content of the soil and whether he needs to irrigate. He can see how much fertilizer he's used in every part of his field. With all that information available, it can be difficult to sort out what is important. "We're finally becoming a data-based decision-making company. Rather than just going out there and having a green thumb and making it happen, you can use research, application rates, and put out different products and become more scientific." To do that, he must be able to figure out what is important. "I can give you a hundred sets of data that would all seem important," Bridgeforth explained, "but we don't have data technicians to analyze the data. You can become engulfed in how much data you receive." Software that uses a rudimentary artificial intelligence system helps with that process. In some instances, the software simply takes the data and decides when to apply more water. Other software saves all the information about application of fertilizer, pesticides, and other relevant data to the cloud, which allows him to refine his management, repeating what worked last year.

The cumulative effect of those refinements results in better crop yields while using less water, fertilizer, and pesticides. "Conservatively, we save 10–15 percent based on how much seed and fertilizer we use, and adjusting those volumes on the fly, rather than just doing the whole field no matter what." As farming becomes more complex, systems that help track and manage these factors give Bridgeforth and other farmers the ability to manage their land in ways that are better for the environment but might be difficult if they had to track every other variable at the same time. The family farm now uses a natural crop rotation, which includes fewer herbicides. Their no-till practices reduce erosion and keep moisture in the soil while saving energy. All of those practices pre-date smartphones and artificial intelligence, to be sure. The ability to add complexity by increasing the usability of information on the farm, however, means farmers can use many techniques that improve management by 1 or 2 percent, rather than trying to choose a few key improvements and focusing only on those.

From the standpoint of looking for ways to reduce the environmental impact of farming, the mundane job of data analysis can be underappreciated. From the standpoint of an outsider, people marvel at the self-driving tractors and combines that do virtually all the work automatically. Those machines are an amazing leap forward, but it is the underlying data that guides how they operate and turns them from technological marvels into machines that contribute to economic prosperity and environmental stewardship.

Microsoft is looking to go far beyond that. Their FarmBeats program combines weather data with soil sensors, drones, TYE (Tethered eYE) helium balloons that monitor the crop, and other small technologies to create maps of the field that aid in applying water and other resources efficiently.

Artificial intelligence doesn't replace the farmer any more than Nest replaces the homeowner in making decisions about what temperature is comfortable in their house. Smartphones and personal technology improve the quality of decisions at the margin, saving a little bit of water or increasing yields above what farmers might achieve using experience alone.

INFORMATION IN THE HANDS OF THOSE WHO CAN BEST USE IT

The combination of personal incentives and artificial intelligence can change the way we apply expertise to environmental problems. Without information about the specifics of a real-world circumstance, subject experts on agriculture, energy, or other complex topics provide useful information that must be translated and applied to local circumstances. No two farms are the same.

Nobody's energy use is the same. There are some very good rules of thumb, and experts can provide guidance that, on average, works adequately for most.

Even the best expert is likely to face hurdles. First, information and expertise that are applicable at a high level can struggle to account for unique preferences of individuals. The "green" building standards I mentioned earlier too frequently failed to account for the local circumstances and ended up using more energy than traditional schools in the same district. Building managers, who dealt with those circumstances every day and had an incentive to reduce energy use, ended up being more effective at finding efficiencies. Even if the recommendations of experts work on average, they don't work for everyone.

Second, experts may not pay a price for failure. Sometimes, they let ideology or personal experience guide decisions rather than the facts on the ground pointing in a different direction.

Neither of these limitations invalidates the value of experts when making public policy or decisions about environmental sustainability. Often, experts have earned the title. Artificial intelligence is able to take complicated information and make it useful to those for whom it matters most. Few people have access to experts to help them manage their electricity use, reduce household water use, or apply the right amount of fertilizer to their crop. Even those who know they could save money may decide the cost of the expertise is greater than the potential savings. Personal technologies using artificial intelligence change that calculation, reducing the cost of information, strengthening the incentives to conserve, and empowering people to take action.

I'm not saying artificial intelligence is more powerful than guilt, but when it comes to helping the environment and combined with the right incentives, AI is starting to give guilt a run for its money.

Aggregating Knowledge
for the Environment

For wind turbines to generate the maximum amount of electricity during their life span, developers need to know the microclimates of potential sites—how and when the wind blows, not just across a region but in very specific locations. Little differences in wind velocity and frequency can change the amount and timing of electricity generation. The US Energy Information Administration reports that costs for wind-generated electricity vary more than other types of energy because turbines are so site dependent.[86] Cliff Mass, a climate scientist at the University of Washington, believes smartphones can help provide an increased level of accuracy when finding locations to site wind turbines for maximum energy generation.

The air pressure sensors inside smartphones are surprisingly accurate. "They are able to tell you what level of the mall you are on," says Mass, by sensing the change in air pressure when you change altitude by just a dozen feet. That pressure data can be extremely useful when predicting weather and the wind potential of a location. As Mass told me, "Pressure is very valuable. Pressure reveals what is happening at the surface and the weight of the atmosphere above you." Mass and his colleague Luke Madaus believe the potential for smartphones as tools to predict weather is tremendous. When they began their research there were between five hundred million and one billion smartphones with the capacity to measure pressure as well as other key pieces of information, including humidity and location.[87] In the same way traffic apps like Waze use real-time information collected from many users to predict the best route to get home, weather apps that read the pressure and humidity data in real time can help predict the weather and understand the microclimates of locations that might not currently have sufficient information.

The pressure data from smartphones might be most useful after wind turbines have already been sited, adjusting to changes in weather that affect wind generation. In a study published in the *Bulletin of the American Meteorological Society*, Madaus and Mass wrote, "Forecasting the positions of fronts and major troughs, even a few hours in advance, can have large value for wind energy prediction since such features often are associated with sudden rapid ramp ups and ramp downs in wind energy generation." The timing of energy generation makes a difference, and being able to predict when more wind-generated electricity will come on the grid can help shift from sources like coal and natural gas to wind. Some energy markets work on five-minute intervals, making even short-term gust predictions valuable.

Creating accurate microclimate forecasts is difficult, and for the smartphone data to be useful it needs to be verified to prevent errors from skewing the information. "The key issue is quality control. That is the key issue with crowdsourced data control," says Mass. How do we know the data are accurate? What factors could be creating a false reading? Unlike temperature, which can be fooled when a phone user goes into an air-conditioned room, pressure is the same inside and outside. "Pressure is better than anything else in terms of collection," says Mass. And it is extremely useful in predicting weather patterns, making it perfect for crowdsourcing. Still, there can be uncertainty, and double-checking the data is important. Mass excitedly notes that, "we had a big breakthrough using machine learning," which allows them to test the data. "We are getting sixteen million pressures a day and we have done some experiments," to be sure the data are accurate and can be verified. Research by Callie McNicholas and Mass using crowdsourced smartphone pressure information found corrected data did better at capturing the structure and evolution of Hurricane Michael, which hit the Caribbean in 2018, than conventional surface pressure networks.[88]

The phrase "big data" is all the rage as computing power and the ability to analyze large amounts of information expand. All that computing power requires reliable information at a usable scale. By collecting pressure data and sharing it in real time, smartphones create a web of sensors filling in gaps where information was scarce. The data they share are immediate and actionable and come at a very low cost. Researchers don't have to pay for or maintain the equipment. They just have to convince enough people to put an app on their phone to have coverage across an area. "Where this may be extremely useful is in third-world countries," says Mass. "They have a lot of smartphones. We might make big improvements in forecasting." Smartphones are increasingly substituting for traditional sensor systems, allowing developing countries to take advantage of smartphone innovation. Being able to predict severe weather a little earlier can improve notification where information is difficult to share.

Aggregating data from a wide variety of sources—data that was not even being collected previously—creates opportunities to address environmental problems that seem intractable now. As valuable as good information is, it also has a cost. In our daily lives we learn to make do with the information we have, relying on mental shortcuts to fill in the gaps in our knowledge. These shortcuts, called "heuristics," are so familiar we don't even realize we are using them. In many cases they work adequately. After trying a few different routes, we drive the same way to work every day, even if we know that on some days traffic will be very bad. Until we had traffic apps on our phones, that was a mostly reliable approach to getting to work on time. Now, we use Waze and Google Maps to see where the traffic is and find the fastest way to get from one place to another. Those apps do for traffic what smartphones could do for weather forecasting and other environmental issues: provide ground-level data in real time that improves decision making.

BRINGING PEOPLE TOGETHER FOR THE ENVIRONMENT

This information isn't created by smartphones. It was already there. The cost of accessing that distributed information, however, was too large to make it useful. Smartphones drive down the cost of collecting and sharing information, helping solve a collective action problem. For researchers or businesses trying to acquire the information they need to make good decisions, the challenge is how to make that data affordable. This is what your grocery store does when they offer discounts for using a loyalty card. They pay you a small amount to make sure you keep shopping there and for the information you provide about shopping habits.

The ability of good information to improve outcomes and create incentives is well known to economists. Nobel Prize winner George Akerlof's work centered on the high cost of what he called "asymmetry of information," where people have access to different amounts and quality of data.[89] The gap between what we know and the information we could use creates inefficiencies. Akerlof noted that when information is limited, sellers of goods have an incentive to offer poor-quality products because buyers are unlikely to be able to detect problems. He used the example of used cars where sellers know more about a car's problems—oil leaks, repairs that need to be made, etc.—than a potential buyer. As a result, buyers worry about purchasing a car that might be worth less than they paid. To counteract this, buyers are cautious, offering less than they otherwise would for a similar car because they price in potential surprises. Sellers of good cars are, in response, less likely to sell their car because finding a buyer willing to pay what a good car is worth is difficult. The lack of information creates a situation that is bad for both buyers and sellers.

Mike Phillips of Sense highlighted the challenge this presents when trying to get enough information so his technology could accurately identify new appliances. "There is a crowd or network effect," to training artificial intelligence to identify appliances. "The only way to make this work is to have good models. The only way to get good models is to have good data. And the only way to have good data is to get them [the Sense monitor] out there," he explained. In order for Sense to effectively identify appliances, which helps users identify where they are using energy, they have to have data from many users. Users, however, may be reluctant to buy a Sense until it can identify many appliances. It costs nothing for people who have a Sense in their house, like me, to share the information and improve the quality of the product. Aggregating the information that is available makes it easier to find ways to save energy.

The web site Zooniverse offers another approach to reducing the cost of gathering information by promoting citizen science. The web site invites people to identify information in photos, adding knowledge that would take a huge amount of time for researchers to gather themselves. The idea began with a project called Galaxy Zoo, using a database of 930,000 photos of galaxies from the Sloan Digital Sky Survey. Researcher Kevin Schawinski wanted to classify each galaxy by its shape, distinguishing between spiral, elliptical, and cluster galaxies. Making these determinations is fairly easy, even for non-astronomers. The instructions on Galaxy Zoo comfort users by saying, "You can answer these questions without any specialist knowledge. Many of the galaxies are distant, so the answer may not always be obvious—just take your best guess."[90] Displaying a photo of the galaxy, the page asks simple questions, like, "Is the galaxy simply smooth and rounded, with no sign of a disk?" In less than a minute, users can add useful information. Galaxy Zoo's breakthrough was to make the cost of adding information very low. The reward for users is the feeling that they have helped improve scientific knowledge and learned something about astronomy. The first message that greets users on the site is: "Few have witnessed what you're about to see." With that kind of enticing introduction, who wouldn't want to spend a few minutes looking at galaxies and answering some simple questions?

The system had tremendous results. When it was launched in 2007, the creators publicized it on BBC and elsewhere and were overwhelmed by the response. The website crashed. As Michael Nielsen wrote in his book about citizen science and innovation, "By the end of the first day more than 70,000 galaxy classifications were being done every *hour*—more than Schawinski had managed" to complete in a week.[91] So many users signed up, each galaxy could be viewed by several volunteers, improving data quality and making it difficult for a single poor classification to skew the data.

The success of Galaxy Zoo spawned Zooniverse, a one-stop site for many crowdsourced projects covering a wide range of topics. Users can choose projects related to climate, like the "Weather Rescue" project where they can transcribe information from weather logs in the UK from the 1860s. As the site notes, there are more than 2.5 million pieces of data on written reports that need to be digitized so they can be useful to researchers. Project organizers claim "this will be the first time that climate scientists and meteorologists from around the globe will have had access to the raw data. This could mean not only that we have a better understanding of the climate from the past, but also could help us understand what the future could look like for us as well."[92]

In the "Nature" category, users can identify pictures with wildlife like sage grouse or listen to seismic recordings to identify potential earthquakes. By listening to and classifying data from seismographs, users improve the quality of the information used to predict future earthquakes.[93] These projects demonstrate the opportunity created by just a small amount of additional information. Collecting that information is now extremely inexpensive. Nielsen wrote, "Online tools offer us a fresh opportunity to improve the way discoveries are made, and opportunity on a scale not seen since the early days of modern science." Whether that is hyperbole or not remains to be seen, but he is certainly right to recognize that information is creating vast new opportunities to make a difference.

AIRBNB FOR BIRDS

The principle of using low-cost data provided by many individuals is also being put to work to expand our understanding of migratory birds and protect the habitat they rely on. Every day, thousands of people head out to bird-watch, adding to their life lists the birds they've seen. Smartphones are now a key part of birders' experiences, helping them identify and catalog the species, time, and location of their sightings. One of the most popular apps is eBird, developed by researchers at the Cornell Lab of Ornithology, allowing birders to store and update their life list, sharing data about the number and location of birds for Cornell's global database. Every day, thousands of sightings are added using the app.

Brian Sullivan, the project leader for eBird, notes that the key to the success of the app was not the data collection but making it a useful tool for bird-watching. The app gives birders what they want: the ability to track sightings and find birds they've never seen before. "When we launched eBird in 2002 we had a notion that people would do it because it was good for birds and good for science, but you have to add a component which is that it is good for [the user]," he says.

Sullivan knows what birders want. He's written several books on birding and other guides and has traveled to Antarctica, the Arctic, Australia, and Central and South America to photograph birds.[94] He realized that what began as a tool to get more information about birds and their locations became an opportunity to connect people with the birds they loved. "We flipped it on its head from typical citizen science where the data is the focus, and now we have a tool that is good for birders [from whom] we then extract the data," he says.

When someone adds a sighting for a particular bird in the app, it uses the GPS tracker in the phone to mark the location and when they saw it. The information is then entered into a database. If anyone makes a mistake, researchers at Cornell have tools that highlight questionable entries to help ensure data quality.

The number of entries is enormous and growing. "We are seeing about ten million observations a month globally, most of them coming from the United States," says Sullivan. That pace is accelerating, and the number of data points grew by about 30 percent per year for twelve years. There is even a world map that displays real-time checklist submissions, with dots appearing as people enter information. Every few seconds, bright yellow dots populate the map, appearing in Tasmania, Colorado, Minnesota, and Baja California in just the minute it took me to write the last two sentences.[95]

eBird began as a tool on the internet where people could transfer the information they collected in the field to a desktop computer. Smartphones made that process much easier. "The ability to enter data from the field has dramatically changed the data volume and the data quality," says Sullivan. "Personally, as soon as we had the mobile app, I never entered a checklist online again. Not only is that convenient but keeping track of numbers of birds in the field was much easier on the phone."

Now the developers are using games to encourage people to fill in data gaps. Birders have their favorite locations, either because the scenery is beautiful, their favorite birds frequent the location, or simply because it is nearby. As a result, Cornell has a huge amount of data in some areas but very little in places that are more remote. So, the eBird team created games to entice people to go places they hadn't been before.

Called Avicaching, the game uses the eBird app and allows users to score points when they go birding where more data is needed. Modeled after geocaching, where people use GPS coordinates and other clues to find prizes, Avicaching rewards users who find birds in locations where data is scarce, reducing what the Cornell team calls "roadside bias." They explain their goal this way:

> Roadside bias is loosely defined as the differences in detectability for certain species as a function of your distance from a road. For example, a Scarlet Tanager is more likely to be encountered in the middle of the forest, and a Gray Catbird is going to be most often found at the edge of a field or in other non-forested habitats. When modeling distribution of bird species, you can greatly improve model results by thoroughly accounting for roadside bias. By collecting data at offroad birding spots for Avicaching, users gather data that will allow Cornell researchers to address these questions in a comprehensive manner.[96]

Birders earn points for each bird checklist they enter for designated locations. The more lists they provide in a week, the more points they earn, and the more data Cornell has to fill in the information gaps. The tangible rewards are pretty simple: T-shirts and baseball caps. For many, however, the joy of birding and being an active participant in citizen science is the reward. Known as "gamification," the competition element of the app provides an additional incentive to make the database more robust and useful. "It is pretty easy to get sucked in," says Sullivan.

Smartphones allow Cornell researchers to dramatically improve the quantity and quality of citizen science. The type of information from birding checklists that used to be hidden in private journals across the country is now being collected and made useful for scientists, conservationists, and amateur birders around the world. eBird users understand they are providing scientifically useful data, but many also see the app as an easy way to make a hobby they love more enjoyable. I was not a birder when I learned about eBird, but thanks to the app I quickly learned to identify the birds in the forest behind my house. It helped me make a connection to the wildlife around me that previously would have required more time and commitment.

Creating that connection is valuable in and of itself, but the team at eBird also used the information to help protect bird habitats in a way that was difficult before smartphones made data entry so easy and immediate.

Partnering with The Nature Conservancy (TNC) in the Central Valley of California in 2013, the eBird team used the data to identify key parcels of land to protect, assisting migratory shorebirds as they move along the Pacific Flyway. "We already had hundreds of thousands of data points to model, which is what the TNC needed to make this conservation action work," Sullivan explained. The data was specific enough, down to about a half-mile of resolution, that TNC could identify farms and rice paddies in the area that would be most valuable during migratory season. Models were created and scientists went to the locations to confirm the model matched the reality on the ground.

Once they had the information, TNC went to the California Rice Commission and to farmers, offering to pay them to create "pop-up habitats" for short periods of time. TNC asked farmers to flood their fields with a few inches of water and let them remain idle during a time when they might otherwise be preparing for the next growing season. TNC ran a reverse auction, asking farmers how much they would have to be paid to participate in the program. Farmers named their price, and TNC rented the land to create the habitat. "If you go to San Francisco, you rent an Airbnb," quips Sullivan. "These birds are moving across landscapes and renting land."

The program, known as BirdReturns, took the eBird data and turned it into real-world bird habitats. TNC's lead scientist for their

California Migratory Bird Program said, "It's been a pretty astonishing success," with 10,000 acres of additional wetland created, and modeling showing "that more than 180,000 waterbirds comprising more than fifty different species used the temporary wetlands—thirty times more than were counted on the dry fields."[97] By renting the land, the cost is dramatically lower than having to purchase it outright. The cost was less than a million dollars to create 10,000 acres of wetland habitat, far less than the nearly a hundred million it could cost to purchase the land. "At these prices, BirdReturns could run for hundreds of years and still cost less than purchasing the land," noted one write-up of the program.[98]

Additionally, turning productive farmland into conservation land permanently can be politically unpopular, as the remaining farmers have more difficulty continuing to work in a community that becomes more detached from a farming lifestyle. Working with the farmers, on the other hand, engages them in conservation and provides financial incentives to compensate for the additional risk and the cost of changing their farming practices temporarily.

Before smartphones, there was no reasonable way to aggregate the information collected by thousands of bird-watchers to act for the environment. Many of the people who contributed data to the eBird database never realized it would end up creating habitat for the birds they loved. They didn't have to. They used the app because it helped them enjoy their hobby and it cost them nothing to share the information with Cornell researchers. Similarly, the cost to create the app and manage the database is vastly lower than paying researchers to do the equivalent field surveys. Thirty years ago, collecting that level of information would have been cost- and time-prohibitive. eBird shows how that reality is changing.

DETECTING POWER OUTAGES IN RURAL GHANA

The same dynamic is helping reduce electrical blackouts. In many parts of the United States, the only way utilities know of a power outage is when customers call them. The United States is slowly replacing old electricity meters with smart meters that provide a signal to utilities indicating the power is off, but this is still not universal.

Some residents of Accra, the capital of Ghana, already have that ability.

Access to electricity is expanding across the globe, and while many communities don't have the infrastructure available in the United States, developing countries are finding ways to use smartphone technology to improve the functionality of their systems without the high costs associated with large capital projects. Necessity is the mother of invention, and smartphones provide new tools to solve these problems.

In the case of Accra, the challenge is reliable electricity. Researchers at the University of California at Berkeley are working to improve reliability by rapidly identifying power outages and the source of the disconnect. Although 95 percent of city residents have access to power, the average customer, according to Catherine Wolfram and Susanna Berkouwer of Cal Berkeley, experienced more than 2,000 hours without power in 2015—about a quarter of a year.[99] Some studies indicate the cost of regular outages reduce long-run economic prosperity by 2–3 percent.[100] As in some cities in the United States, it can be a challenge to determine where the power is out and isolate the problem. The team at Cal Berkeley developed an app to help solve that problem.

Called DumsorWatch ("dumsor" means "off and on" in the Akan language), the app tracked the charge state of the phone and whether the phone was moving. It also tracked whether the phone was connected to a wifi network. Using these pieces of information, the app determined if there was a power outage. The researchers explained, "Phones record their charge state, so if a phone is plugged in and then suddenly stops charging, there might be an electricity outage. This is more likely an outage if the phone didn't also start moving, as movement would suggest someone might have unplugged it and put it in their pocket." Additionally, if the phone simultaneously disconnected from wifi, that was another indication that power was out and the wifi router also lost power. When it detected these changes, the app sent an alert using the cell service, warning the utility that there might be an outage. If the utility saw several signals appear all in the same area, it was a good indication that the power was out in that neighborhood, and it helped pinpoint where the disconnect occurred.

To test the theory, researchers ran a pilot with about two thousand people in one part of the city. The result? It worked as expected. The "improved information will help quickly and accurately assess the extent of an outage. That way, a utility can dispatch repair crews where they're most needed to better prioritize these limited resources." It also means electricity providers can spend their resources on expanding access and improving reliability rather than on additional infrastructure that might be necessary to track outages.

Like the smartphone app tracking air pressure, DumsorWatch automatically collected data provided by a connected smartphone without user input. It simply aggregated data that would not be available without significant cost. In the United States, some utilities delayed installing outage-detecting smart meters due to the high cost—a price that was even more unthinkable in Ghana. The app, by way of comparison, reduced the cost of the information to almost zero. Aggregating that information helps utilities fix outages, returning service so the utility can earn the revenue necessary to expand service. As the researchers note, "The quality of electricity was one of the most hotly debated topics during the 2016 Presidential Election in Ghana," so a simple app that improves electricity reliability could also improve political stability. And this is all possible simply by adding an app and plugging in a smartphone.

The success of the initial pilot led the researchers to create a dedicated technology called PowerWatch that plugs into the wall and reports outages directly.[101] The device uses less than 10 percent of the electricity of charging a cell phone and it collects data that is useful to identify outages and to improve overall reliability of the grid. The technology is also being used in other countries. The smartphone app became the low-cost test for the hypothesis, providing the results that made possible the next step to improve reliable access to electricity.

DONATING DATA TO IMPROVE ENERGY EFFICIENCY

In North America, we are blessed to have a mostly reliable electricity supply. Instead, the focus of homeowners in developed countries like the United States and Canada is to use every kilowatt hour in the most effective and efficient way. It saves money and helps the environment. How to be efficient is not always so simple or obvious. Is my home energy efficient? Should I spend money on more insulation for my house? Answering these questions requires more analysis and information.

Making buildings more energy efficient is an important part of reducing the environmental impact of energy. The Energy Information Administration notes that about 40 percent of US energy consumption is associated with residential and commercial buildings.[102] Understanding this, environmental researchers have tried to find ways to improve building efficiency. The US Green Building Council, the Living Building Challenge, and other systems have been designed to reduce the amount of energy used in buildings. One of the central pieces of the first "Green New Deal" was to "upgrade or replace every building in the US for state-of-the-art energy efficiency."[103] These efforts, however, have big problems. First, they are extremely expensive. The Living Building Challenge is designed to push the limits of building efficiency, but it comes with a high cost and is not applicable for the vast majority of buildings. Second, these systems are cookie-cutter approaches that are difficult to apply effectively.

Creating effective solutions requires approaches that are customized to particular buildings. That can be expensive or impossible without information to determine why a particular building is losing energy. "Despite the importance and universality of building energy use," wrote Meta Brown, a journalist who focuses on data analytics, "research on the details has been surprisingly limited because the cost of data collection typically restricts studies to small sample sizes and little or no information about what's going on within buildings."[104] Ecobee, a company that has a smart thermostat that competes with Google Nest, is collecting the data to answer exactly those questions.

I mentioned Ecobee in the last chapter, highlighting the effectiveness of the artificial intelligence in their thermostat at saving energy. The company is also using data provided voluntarily by its users to improve the performance of their product and learn more about how to make homes efficient. This involves increasing the number of sensors used by the thermostat and making that information available to users. "We're on the verge of being a smartphone company," said Fatima Crerar, who served as Ecobee's director of social impact and sustainability. "We've taken our success in the smart thermostat and moved into smartphones." Ecobee says its users are already saving up to 23 percent of their energy use. With additional information and some analysis, they believe that percentage can go even higher. Combining artificial intelligence with improved data is key to taking that next step. Crerar explained her vision for how Ecobee could help. "People have to change their behavior, but AI takes all of that away. Ecobee has the goal of taking chaos away in the household, so people can live life."

Most thermostats collect temperature information at one location—the thermostat itself. This can be a fine approximation, but there can also be big differences depending on whether it is upstairs or downstairs and other factors specific to a house. Ecobee added more sensors to help the system calibrate energy use and temperatures. "Ecobee 3 was the first smart thermostat that came with a sensor," said Crerar. "We are actually trying to solve a problem. The thermostat is in one spot and measuring the temperature in one spot." Adding sensors improves the accuracy of the thermostat, but it "also increases the intelligence of the thermostat so it can deliver comfort and savings more effectively," noted Crerar. The information gathered by those sensors can be used for more than just keeping your house at a comfortable temperature. It can be aggregated to learn more about how homes can become more efficient and reduce the environmental costs associated with energy use.

Ecobee launched a program called "Donate Your Data" that asks users to share the information from their thermostat and sensors so researchers can improve their understanding of home energy use. "Sharing anonymized data from your Ecobee smart thermostat can help scientists advance the way to a sustainable future," says the web page encouraging users to sign up for the program. In addition to the readings provided by the thermostat, Ecobee asks for information that provides context about the home and how it is used. When they sign up, users provide the city and state or province, house size, number of floors, estimated home age, and number of occupants. The system adds heat and cooling temperature settings, occupancy schedules, indoor and outdoor temperature and humidity, and HVAC run times. "We have data in five-minute intervals," says Crerar. "The data is high fidelity that has all the sensor information." When I spoke with her, Crerar said they had already signed up more than 31,000 users. With that many volunteers, the data set offers some real insights. "It is a large data set. Corner to corner on the continent. Different climatic regions, different sizes of homes, and different people."

Many people are willing to share data if they think it will help us become better stewards of the planet. To make people comfortable, Ecobee is careful to keep the data anonymous. "The assumption we had is that people are terrified," said Crerar, "but when they see the benefits and the scale of the problem, and they can volunteer, people will say yes. It is how you ask and what commitments you make. We have been transparent, and we just ask people to opt in. We are transparent about who is getting the data and why." Participants are rewarded in two ways.

First, they get to be part of something important. Just as Galaxy Zoo entices users by promising they will see things that few other people have witnessed, Ecobee tells people their contributions can help find solutions to reducing the impact of CO_2 on our climate. Ecobee shares the science and research created by their data. That research becomes a tangible connection between the decision to participate and helping the environment. "We want them to feel proud of what they are helping out with but also hopeful," Crerar explained. Users can see that they are part of the solution.

Second, users can benefit, at least indirectly, from some of the research. One project by researchers at Cornell University used Ecobee data to design a simplistic energy audit. It uses sensor data to see how well a home holds a temperature once the furnace turns off. Cornell professor Dr. Howard Chong analyzed the data to see why homes vary so much in their energy efficiency. "The Donate Your Data set is confirming that it's also whole home performance leading to those variations, not just behavior," he explained.[105] Chong developed a "leaky house test" that produces a score for your home. Users set the room to a comfortable temperature and then turn off the thermostat. If the temperature drops quickly, the home gets a poor score. If the temperature stays constant, your home earns a good score. That single piece of information allows homeowners to decide whether spending money on additional insulation is worthwhile.

In addition to Cornell, the data has been used by the National Renewable Energy Laboratory, University of California Davis, the Government of Canada, the University of Oklahoma, the American Council for an Energy Efficient Economy, and several others.[106] Collecting that amount of information without the smart thermostat would have been impossible. By way of comparison, a study of government-sponsored energy efficiency programs by Michael Greenstone, a former economic adviser to President Obama, completed interviews with 2,700 homeowners.[107] Each of those interviews was probably more insightful than a simple data stream from a home enrolled in Donate Your Data. The Ecobee program, however, is continuous and has ten times the number of homes. Each type of research has a role, but being able to collect environmental and energy data from such a large number of homes opens new avenues that could take a big bite out of the energy used by buildings, without the massive price tag of the Green New Deal or other politically centered programs.

All of that without users having to do anything more than set their thermostat.

HOW HACKERS ARE SAVING FISH

The ability to aggregate data from many sources can also reduce overfishing. As a child in Jakarta, Natasya grew up loving the ocean. I met Natasya in Seattle at the Fishackathon, a worldwide competition among app developers to create programs that help improve the sustainability of fishing across the globe. She was enrolled in a community college and studying computer science. Only in her early twenties, she organized the Seattle event where about fifty local programmers spent the weekend tackling fisheries problems.

For her, combining her programming skills with concern for the ocean and fisheries was natural. "I want to be a software engineer and I love the oceans and love to snorkel in Indonesia," she said. "I see lots of troubles about stealing fish. Knowing this involves fisheries was pretty exciting." The fact that a woman from Indonesia can organize a hackathon in Seattle to solve problems in the Indian Ocean and Africa by providing fishers with simple technology encapsulates the opportunity the world has to address environmental problems in remarkable new ways.

The Fishackathon was built around the idea that collecting and applying information can solve difficult problems. Sponsored by Hackernest along with the US State Department and the Government of Canada, the event brought 3,500 participants together in forty different cities. Over the weekend, the "hackers" worked to solve problems identified by organizations looking to create sustainable fisheries and improve the lives of fishers around the world. Seattle was joined by cities as distant as Perth, Australia, as well as Benin City, Nigeria, and Toronto, Canada.

The Fishackathon's mission is to bring together "thousands of concerned designers, developers, and subject matter experts for a weekend to build practical tech solutions to endemic problems defined by the world's most respected fisheries experts."[108] Those experts include the Environmental Defense Fund and the US Geological Survey, who create problem statements that guide programmers in developing solutions. Individuals then form teams and discuss the problems that sound most interesting. When I arrived at the Seattle location to mentor the teams, there were about thirty programmers in the room, and leaders created teams based on their interest in various challenge sets. In addition to the altruism and camaraderie offered by the Fishackathon, teams were also competing for cash prizes for the best solution in each city and worldwide. After all, as much as people want to help the environment, it does not hurt to add an incentive.

The problems are designed so programmers can either solve them or get a good start in the course of a weekend. Hackernest CEO Shaharris, who organized the first few Fishackathons, told me they interview those looking to offer problem statements to make sure they aren't too broad and focus on the key elements of a fisheries challenge. "The reason hackathons work is that you are forced to take the problem and whittle away at it and cut it down to a binary problem," he told me. "The time limit forces an achievable goal and forces you to really plan and strategize." The goal at the end of the weekend is to have a prototype or proof of concept that can go somewhere. "Innovation is taking an idea and making it practically applicable," he said.

The problems facing the developers were pretty diverse.

In 2016, participants were challenged to create an app that could identify and locate "ghost nets"—fishing gear lost at sea that continues to float and ensnare fish, marine mammals, and other sea life. In its problem statement, the Global Ghost Gear Initiative (GGGI) noted, "Hundreds of thousands of whales, dolphins, seals, and sea turtles are killed each year due to entrapment in lost gear, and their migratory nature and capacity for lost gear to drift cause entanglement on a global scale."[109] GGGI requested an app that would allow individuals to report ghost gear where they found it, take a photo, and provide identifying information to help governments and organizations track the gear down and remove it from the water. The problem statement offered several sources of data and even requested that the app help identify the source of the gear by using information about currents and the type of gear used in different parts of the world.

In 2018, one of the Seattle teams selected a problem statement developed by the American Fisheries Society. Like the ghost gear challenge, the purpose was to allow individuals to report environmental hazards. "Monitoring is inadequate in most states due to limited funding and lack of infrastructure," the challenge statement noted.[110] "Furthermore, many conditions that affect aquatic health, fishing quality, and environmental quality are difficult or impossible to measure." Even the most dedicated and well-funded government environmental agencies can't be everywhere. Monitoring and enforcement are always difficult, especially for small and distributed problems. Large pollution outfalls pouring into a river are easy to identify and monitor. Algae blooms caused by fertilizer runoff from someone's lawn are virtually impossible to track. Using crowdsourced data to fill that monitoring gap is precisely what the Fisheries Society had in mind, noting, "Resource managers (e.g., the state Department of Natural Resources) lack the bandwidth to continually monitor and measure these changing conditions even though the information would be very useful for anglers and others engaging in freshwater recreational activities." The challenge was to create a crowdsourced database of environmental problems in bodies of water. The Fisheries Society requested "an easy-to-use, easy-to-share application for anglers and other members of the public that records a variety of user-submitted environmental conditions and displays them spatially for users." Such an application "would help tremendously in protecting freshwater ecosystems. The creation of this application in a manner which can be easily adopted by a resource agency would help them keep track of conditions and quickly and effectively respond to issues that arise." The app would allow recreational and commercial anglers to identify the location of the problem on a map, giving them the opportunity to mark fish kills, algae blooms, and other types of pollution.

The Seattle team created a demonstration project called H2Flow that allowed the user to mark what they found by turning identifying environmental problems into a game. Like a scavenger hunt, the app would reward users who reported a problem with tokens. In the same way that eBird uses gamification to reward bird-watchers for logging reports, H2Flow was built with campers and scouts in mind—groups that would enjoy the challenge of finding environmental problems and would appreciate recognition for their efforts. Timothy Lowry, the H2Flow team leader, said the goal of the app was to put "the citizen closer to the [environmental] agency through data exchange."

Crowdsourcing information is useful, but accessing new information is most valuable when it informs decision making, providing incentives to use resources more wisely. That's what the Environmental Defense Fund wanted to achieve with its problem set directed at helping fishers get the best price for their catch. In many parts of the world, fishers have a difficult time getting good market information. While at sea, they have several choices of ports where they can put in and sell their catch. Choosing the right one—where demand and prices are highest for the type of fish they caught—can be hit-or-miss. Fishing vessels can't go from port to port looking for the best deal. That takes fuel and time they don't have. As the problem statement noted, "Fishers want to maximize their catch by selling it off as soon as possible (for freshness), which makes it difficult to spend time evaluating prices to choose buyers." Spend too long looking for the best deal and fishers could be stuck with a boat full of spoiled fish.

The EDF asked programmers at the Fishackathon to develop "a platform that allows fishers to report 'received' prices and buyers to report 'offered' prices in real time."[111] Each time a fisher puts in at a port, they would report the price they received, signaling others where markets are good and the ports to avoid. By making this information available on the app it "would optimize profitability and increase market transparency, leading to a more competitive, fairer marketplace."

The inspiration for the app came from EDF's work in Mexico. Tim Fitzgerald served as EDF's director of impact for the Fishery Solutions Center and saw how market information helped fishers become more financially stable, allowing them to focus on sustainability. "In Mexico, there are lots of independent small-scale fishermen catching what they can and coming back to the beach and selling it to the same person and accepting whatever price they are offered," he explained. "They may not have or know about other options, but we thought—we've done some pilot work in Mexico—if there was a way for fishermen to have a better understanding of the market around them, not just the one person, maybe that changes the dynamic and gives them more of an upper hand." This financial stability gives fishers the opportunity to make decisions that consider the sustainability of their fishing practices rather than just trying to catch everything they can and be happy with the little money they make. As Fitzgerald noted, "If all they've ever known is catch as much as you can and sell it for what you can, you can't have a real sustainability conversation in that environment." Jack Welch, the legendary former CEO of

General Electric, put it this way: "You can't grow long-term if you can't eat short-term." Similarly, you can't manage a fishery for the long term if you are struggling to feed your family every day.

EDF asked programmers to develop an app that would help fishers maximize their income and reduce the number of fish that spoiled in their boat while they were looking for buyers. There are other apps that do something similar, but the challenge was to develop a solution that accommodates the technology available to poor fishermen. "Almost everything was smartphone focused, which is fine, but the challenges are directed at small-scale fisheries, fishers, workers," said Fitzgerald. "There is a tendency for the coders to go the hard-core, smartphone app route." The team that developed the winning version of the app, called Finnder, developed the system to run on flip phones. They noted that at the time, about 70 percent of the fishers only had a flip phone.[112] The system allowed users to text the information they wanted, such as a list of prices at different ports, and the system would send a text back to the phone.

Organizations like EDF then choose the best solutions to their Fishackathon problem statement and work with the team to develop a complete, workable version of the app. Developing a complete solution in just forty-eight hours is rare, so there is often additional work required to finish the project. The winning teams have the incentive of a prize for the best app in each city, and then for overall winners, the opportunity to create a business around the solution they've developed. Even for those who do not win, there is a sense of accomplishment. One participant at the Seattle location where I was, told me, "You feel like you are helping. It doesn't feel like a waste of time."

The Fishackathon also highlights the value of nonpolitical approaches to addressing environmental problems. Initially, the State Department sponsored the event during the Obama administration. When Donald Trump became president, some in the State Department worried that the project would be sidelined.

To avoid the political concerns, they gave Hackernest—a private organization based in Canada—control of the event. Hackernest CEO Shaharris told me, "With the change in administration, we had no certainty. They asked Hackernest to take it on," with the State Department as a partner rather than the lead. "They wanted to give us a grant, but we wanted them as a partner so everyone is accountable. Being a partner with the US State Department gives us access to the embassies." Since the Fishackathon addresses problems across the globe, Hackernest wanted to have the participation of governments in the countries where events were being held. Compare the response of an embassy receiving a call from the US Department of State to one from Hackernest. Which would they call back? Having the State Department involved was a key part of making the event successful, but the goal was to remove the politics and focus on the results. "Hackernest is apolitical," says Shaharris. "We don't care too much about the governments or which administration we are working with. Specifically, for Fishackathon there is no political motive. It is a logistical thing."

Watching the programmers choose a problem, organize a response, and then develop the solution was really enjoyable. It connected programmers, who might otherwise feel distant from environmental solutions, to real-world problems and people that could have a meaningful impact on sustainability. The apps they developed had the potential to do the same for others using the solutions they developed. It is a great example of the power of smartphones to aggregate information and connect people to the environment.

There is an interesting postscript to the story. After 2018, Hackernest decided to stop running the Fishackathon. Despite the promise of the solutions that were created, they argued hackathons weren't having the effect they wanted. Shaharris outlined some of the problems he saw in a February 2019 blog post.[113] His judgment was pretty brutal.

Among his complaints, he said that "hackathons don't solve important or persistent problems." The limits of a forty-eight-hour hackathon make it difficult to come up with solutions that match the complexity of many of the problems facing sustainability. "Solving big, endemic problems takes time, effort, and understanding. That won't fit into a three-minute demo," he noted.

He also complained that many of the hackathons were "wasteful, inefficient, and demoralizing." The math on this point is fairly compelling.

A big hackathon may have 300 participants working on complex problems all weekend. That's 14,400 hours of high-functioning human effort. Let's say 5 percent win prizes and praise. To the other 285 participants, the hackathon basically says, "Thanks, you weren't good enough to win. The 13,680 hours of hard work, effort, and good intentions you poured into the weekend amount to little more than a fun learning experience. Go home."

Finally, he says it can be very difficult to judge the merits of each solution when teams are only given a few minutes to present their solution. "The one- to three-minute demo format makes it a contest of design and presentation skills, not technical prowess," he laments. "Hackathons are won by designers and charming presenters; teams can win by making pretty mock-ups with little actual code because they know nobody looks under the hood." I can testify personally that this is the case. I wasn't a judge, but I watched the presentations, and it was difficult to know what really worked. Indeed, one team claimed they had developed a blockchain data system as part of their solution but admitted to me a couple weeks later that it didn't actually work. Without looking at the actual code, the judges could never catch that problem.

Hackernest isn't just folding up its tent and going away, however. Shaharris and his team developed an alternative system called

PRIME that they hope will address these and other concerns, engaging the creativity of developers to solve problems in a way that is more likely to create success.

Initially when Shaharris shared his concerns with me, I was a bit crestfallen. The Fishackathon had been enjoyable, and it was fun to watch people using their skills to address problems I cared about. Ultimately, however, I got excited about this new development. The reason smartphones and small personal technologies hold such promise for environmental solutions is that they create opportunities for dynamism. As I noted earlier, what worked in 1970 doesn't necessarily work today. Solutions have to evolve. The nature of the environmental problems we face is changing; so must the solutions. That mindset is at the center of the smartphone environmental approach, and using local knowledge and incentives to find better solutions is the way we are going to tackle these difficult environmental problems. The more I thought about it, the more I loved that Hackernest was so brutally honest about the hackathon's shortcomings. Much of my work in environmental policy laments the fact that people who care about the environment become emotionally attached to solutions because it makes them feel good even if the policies don't work. I realized that I shouldn't make the same mistake and become emotionally attached to hackathons, even if it felt good to be part of it.

That hasn't stopped others from using hackathons. Nor should it. Some groups find them valuable. A conservation technology NGO called **WILD**LABS, on the other hand, is connecting conservationists directly with programmers in the hope that collaboration will result in more robust and usable solutions.

Smartphones and connected technologies are rapidly evolving to create new solutions that overcome the limits of information. Their ability to take small bits of data—such as whether a phone suddenly stopped charging—and aggregate them into usable information is pretty remarkable. The phone in your pocket can help wind turbines become more efficient by sharing local barometric pressure. It can tell fishers where they can get the best price for their catch, ensuring they aren't stuck with a boat full of spoiled fish. These don't seem groundbreaking, and yet that may be the most remarkable part of the story. It is a testament to how much smartphones have changed our world that what was impossible less than a decade ago now seems mundane. By radically reducing the cost of information and collecting it in a usable fashion, smartphones are opening vast new opportunities to turn information into environmental stewardship.

How Diversity Removes
Thousands of Cars
from the Road

There are few better examples of the past success of government-led environmental policy and the need to embrace new technological solutions than transportation-related air pollution. It is one of the major success stories of government regulation. Highway emissions of nitrogen oxides, a family of pollutants associated with smog, has declined by 80 percent since 1970 thanks in part to that year's Clean Air Act and the introduction of catalytic converters in cars.[114] This occurred even as the number of miles traveled by drivers in the United States has more than tripled.[115] This combination of increased mobility and pollution reduction offers the best argument for top-down environmental policy.

Unfortunately, while a catalytic converter can control the pollutants that cause smog, there is no way to add a device to a car that reduces CO_2 emissions. This problem requires different solutions. The list of options to reduce transportation-related CO_2 emissions has been limited, focusing on increasing gas mileage with hybrid cars, increasing the number of electric vehicles, and moving people into public transportation.

For urban planners, creating a transportation system that reliably provides mobility is extremely complex. Alain Bertaud, who worked as the principal urban planner for the World Bank, outlines the stakes of transportation planning. "Failure to manage urban transportation in a manner that maintains mobility results in congestion," Bertaud wrote in his book *Order without Design*.[116] He explains that "Congestion has a dual negative effect: It acts as a tax on productivity by tying down people and goods, and it degrades the environment and increases greenhouse gas emissions." While the stakes are high, the solution is elusive.

Public transportation—buses, trolleys, light rail, subways, ferries, etc.—has been a major focus of climate activists looking to reduce transportation-related CO_2 emissions. Project Drawdown, which developed a list of projects to cut greenhouse gases, notes, "Public transit can keep car use to a minimum and avert greenhouse gases."[117] In addition to being less carbon-intensive than other forms of transportation, it also reduces the costs of maintaining a car, the time wasted in traffic, the time spent finding a place to park for every trip, and the cost of paying for parking. Those costs are incurred for a vehicle that's typically idle for most of the rest of the day. That underutilization means there are more cars than are necessary to supply the transportation needs of their drivers. A car sitting in a parking lot or garage or on the street wastes space and represents a resource that could be put to better use. Still, the US Census reported the share of employees using public transportation in 2018 fell to below 5 percent, a level that had been fairly consistent for more than a decade.[118,119] Meanwhile, the census data showed that the number of people working from home increased, surpassing the number of people using mass transit even before COVID-19 encouraged people to telecommute.

Public transportation as it is currently designed is inflexible, which is why people frequently choose other options. Bertaud notes that a significant percentage of trips are "chained trips," which combine several stops—dropping children off at school and then going to work or shopping. "While chained trips are transport efficient, they are nearly incompatible with public transit or carpooling," he writes. No matter how good public transportation is in theory, if it doesn't serve riders well, or requires unsustainable amounts of funding, it isn't effective. An environmentally friendly choice that nobody chooses doesn't help the environment. Some argue bike commuting is an effective way to cut CO_2 emissions, but only about one-half of 1 percent of people bike to work.[120] How can we combine the resource efficiency of public transportation with the usability of personal vehicles? To replicate the success America saw in reducing air pollution during the past fifty years while preserving mobility will require a diverse set of new options.

Providing a diversity of choices increases the chance that we will find approaches that are effective and appealing to people trying to balance their lifestyle with the desire to cut their environmental footprint. Personal technology and access to information have increased the ability to experiment with transportation options, increasing the diversity of choices. And rather than putting all our eggs into one basket, diversity makes environmental progress more durable and immune to shifts in politics or economics.

SMART TRANSPORTATION OPTIONS

A growing diversity of transportation choices, made possible by smartphones, is a prime example of these new environmental options.

Ride-sharing services like Uber and Lyft, founded in 2011 and 2013, respectively, supplement and, for some commuters, replace public transportation. "If you're commuting home in Los Angeles on the bus, that means waiting an average twenty minutes or more at a stop," noted one article.[121] "Uber, by contrast, has an average wait time of five minutes." One transit rider in Philadelphia said, "If I get out on time, I take the bus. If I'm late, I'll take Uber or Lyft."[122]

These services quickly added options for sharing rides with other passengers for a reduced fee, a service no traditional taxicab offers. Uber launched uberPool, matching riders who are going in a similar direction and allowing them to share the cost of the Uber ride.[123] It takes a little longer than standard Uber rides, but if you have time, you can save money.

What about a way to use cars that would sit unused for most of the day, and put them to work? A company called Car2Go did just that.

In 2011, Car2Go parked a fleet of two-seat smart cars around Seattle. Using a smartphone app, riders could find the Car2Go vehicle nearest them, unlock it with their phone, and drive it to their destination, paying for the time they used the car. Renting a car wasn't a new idea, but the ability to use one's smartphone to find a car, unlock it, and pay for it made it possible to go from renting a car for a day or more to less than an hour. With rates of about fifty cents a minute, it was still significantly less expensive than a ride from Uber or Lyft for the same period of time.[124] The smartphone made it possible to take the time that cars would just be sitting in a parking space and put them to use.

Car2Go wasn't the only company to offer a car-sharing service. Zipcar launched in 2000, before smartphones were available, and initially offered a service that provided short-term car rentals. Smartphones made the process much easier. Purchased by Avis in 2013, Zipcar moved to a hybrid of the traditional car-rental model, offering cars for several days and smartphone-based hourly rentals. Unlike Car2Go, which offered one-way rentals, Zipcar used a round-trip model. A one-hour rental cost as little as nine dollars, and Zipcar even paid for the gas.

Cities like Seattle saw positive results from the growth of car sharing. A survey by researchers Elliot Martin and Susan Shaheen at the University of California at Berkeley found that some who used car-sharing programs like Car2Go decided they didn't need their own cars at all:[125] 14 percent of car-share members in Seattle indicated that they gave up a vehicle after joining the service.[126] Half of those said this was due, in part, to the availability of a car-sharing program. The researchers extrapolated that there were about 4,500 fewer cars on Seattle streets as a result of programs like Car2Go. A three-year study published in 2016 found similar results when the research was expanded to include San Diego; Washington, DC; Calgary; and Vancouver.[127] The survey of 9,500 Car2Go users revealed that an estimated 28,000 privately owned cars were kept off the streets, up to 5 percent of people sold their cars, and up to 10 percent decided not to purchase a car.

Zipcar claimed similar results. In their 2018 impact report, they stated that their 12,000-vehicle fleet kept 156,000 cars off the streets nationwide.[128] These numbers are based on a study from 2010, so a degree of skepticism is warranted, as with any corporate sustainability report. Adam Cohen, who researches car sharing at the Transportation Sustainability Research Center at the University of California at Berkeley, notes that while this number is at the top of the range identified in several studies, it is not unreasonable.[129] "This is one of the areas that has been very extensively studied," he told me. "The methods vary slightly, but they generally tend to show similar numbers." Even if the real number is one-quarter of what they claim, Zipcar is making a positive environmental contribution.

Martin and Shaheen found that some people had used Car2Go the same way they would have a personal car: to take incidental trips as well as for commuting to and from work. This leads some to wonder if car sharing is actually reducing environmental impact. Researchers Regina Clewlow and Gouri Mishra at the University of California at Davis found that many car-sharing trips would not have been taken otherwise. They reported that "49 percent to 61 percent of ride-hailing trips either wouldn't have been made at all if these apps didn't exist, or would have been made by foot, biking, or transit."[130] That may lead people to think car sharing is bad for the environment.

That concern, among others, has induced some city politicians to argue that ride sharing and car sharing reduce transit ridership or undermine the social goals they hope to achieve. Real-world experience has demonstrated that eliminating options like ride sharing ends up being worse for the environment. After Austin severely restricted ride sharing, causing Uber and Lyft to pull out of the city, there was evidence that the number of personal car trips increased. One study found that 9 percent of Austin drivers surveyed said they bought a car to make up for the lack of ride sharing.[131] That change increased the costs to drivers and their environmental footprint.

The same is true for car sharing. On the whole, Martin and Shaheen found the net effect of car sharing was environmentally positive. Drivers who sold their car or didn't buy one used so much less fuel that their net impact more than offset that of those who used car sharing for extra trips that would not have been made otherwise. As a result, car sharing reduced total greenhouse gas emissions. Car2Go also reduced, on average, the total number of miles driven by its members. The fuel efficiency of the smart cars—about 40 mpg in the city—meant each mile used less gasoline than the cars they replaced. In Calgary, this meant a 4 percent reduction in CO_2 emissions for each household using Car2Go. In Washington, DC, the reduction was an amazing 18 percent. Zipcar's research tells a similar story: their users self-report driving 40 percent fewer miles than they did previously.[132] Many use car sharing and ride sharing because they are convenient, inexpensive, and consumer-friendly. The environmental benefits of car sharing and ride sharing are a happy side effect that may make these options more attractive to potential users, but those willing to build their life around reducing environmental impact were probably already taking public transportation, riding a bicycle, or walking. For those who want more mobility than those options provide, car sharing provides the choice of a personal vehicle when circumstances require one. Zipcar notes the average trip in their cars is forty-seven miles.[133] Rather than use car sharing around town, users choose it when they need to take longer trips where public transportation would be more complicated or time-consuming—like chained trips. For other trips users can choose options that are faster, less expensive, and environmentally friendly. The benefit of car sharing isn't necessarily that each individual trip itself is more environmentally friendly. The availability of car sharing means users can choose a lifestyle that doesn't require them to buy a car and can use lower-impact options.

A DIVERSITY OF SOLUTIONS IS GOOD FOR THE ENVIRONMENT

The diversity of transportation options we now enjoy makes it more likely that the environmental benefits we are beginning to see are sustainable and will grow. University of Michigan professor Scott Page, whose book *The Choice* focuses on how diversity is key to solving big problems, says it is critical to provide options and the ability to act. He observes that "diverse people working together on a problem they care about with a sense of agency" are more likely to find workable solutions. Each person sees problems in ways that are influenced by their unique experiences and needs. "A vegan categorizes restaurants differently than an omnivore does, and environmentalists and capitalists see new nuclear power plants through different lenses," notes Page.[134] "Environmental capitalists see them differently still."

The diversity of perspectives offers two advantages to finding effective solutions. First, it can make individual solutions better. Individuals with diverse experiences, perspectives, and information working together to address a problem help avoid blind spots that can doom solutions imposed from small groups of "experts." Nobel Prize winner Daniel Kahneman argues that experts, for all their skill, can suffer from a narrowness of vision. That myopia can lead small groups to commit to solutions that fit their worldview but don't address the full range of demands influencing the problem. This mindset is captured in the saying that to a hammer, every problem looks like a nail. "We know that people can maintain an unshakable faith in any proposition, however absurd," says Kahneman, "when they are sustained by a community of like-minded believers."[135] What sounded great in a small room or looked good on paper ends up failing in the real world. In Kahneman's research, he found that "experts resisted admitting that they had been wrong, and when they were compelled to admit error, they had a large collection of excuses: they had been wrong only in their timing, an unforeseeable event had intervened, or they had been wrong but for the right reasons. Experts are just human in the end. They are dazzled by their own brilliance and hate to be wrong."[136]

It isn't that experts aren't smart. The challenge is that one-size-fits-all approaches no longer succeed in our diverse, increasingly interconnected world. Asking experts to find one or even a few large solutions is unrealistic. The key to solving problems is to create an environment where many approaches emerge, and participants choose the ones that work best. As Page notes, "Knowledge is distributed. The only way to leverage all that knowledge is by having diverse sets of people involved."

Second, the low cost of innovation allows the creation of a wide diversity of solutions that target specific audiences. Although each option caters only to a subset of people, that focus allows them to be more effective individually and more impactful in the aggregate. This allows niche products to emerge that would not be possible if guided solely by government agencies trying to implement one-size-fits-all solutions (or politically expedient ones). As Matt Ridley, a British science and economics journalist, writes, "Political decisions are by definition, monopolistic, disen- franchising and despotically majoritarian." By way of contrast,

Markets are good at supplying minority needs. The other day I bought a device for attaching a fly-fishing rod to my car.... Thanks to the internet, the economy is getting better and better at meeting the desires of [small groups]. Because the very few people in the world who need fishing rod attachments…can now find suppliers on the web, niches are thriving.[137]

What the internet and smartphones have done for fly-fishing accoutrements, they are beginning to do for environmentally friendly transportation.

Diverse needs can now be met as never before, which helps create communities and provide instant response. Uber, for example, now offers many different kinds of services around the world, tailored to specific needs. In Nairobi, Uber added a service called Chapchap, which is designed for short, fast trips ("chapchap" means "faster" in Swahili) within the city.[138] The flexibility of technology allows for rapid experimentation.

In fact, that growing diversity of transportation options and the competition it created played a role in the decision by Car2Go and the BMW-owned car-sharing company ReachNow (which merged in 2018 to form Share Now[139]) to leave the United States as of February 2020.[140] They made the decision for a number of reasons, including "a rapidly evolving competitive mobility landscape" and their desire to "focus on the remaining seventeen European cities" where they provide service. This shift is part of the process of finding solutions that are both environmentally effective and economically sustainable. Not every innovation will be worthwhile, or work for every situation, and failure is part of the process of identifying solutions.

Waze, a Google-owned app that shares real-time crowd-sourced traffic and other navigation information, also diversified. Waze Carpool enables drivers to offer rides to people nearby who are headed in your direction. In addition to saving fuel costs, this is a benefit in cities with special express lanes on freeways for high-occupancy vehicles (HOVs).

While offering a diversity of solutions increases the chance of finding a great transportation option, sometimes a little more is necessary. A 2021 study of ride-sharing options for the US Department of Transportation found that "making mobility options available, while essential to behavior change, does not mean that travelers will change their behavior immediately or at all. What is missing is actively presenting the mobility options to commuters and engaging them through appropriate means (e.g., gamification and incentives) in order to trigger the desired behavior changes."[141] There are options that take things that one step further by actually paying people to carpool.

Partnering with large companies who want to make commuting easier for their employees, Mark Cleveland created an app called Hytch to reward people for carpooling. Using GPS, the app can tell if people are riding together and pays them by the mile. In some cities, large employers are required to reduce the number of commuters; partnering with Hytch allows companies to track people who carpool and provides an incentive for the employees.

It may even prevent a few fistfights. Parking is at a premium at some companies and there is serious competition for the best spots, which, Cleveland explained, has resulted in more than one fight in the parking lot of one of his corporate sponsors. With Hytch, the number of cars competing for those spots is reduced. One of Hytch's clients even paid them to reward employees who carpooled to locations with limited parking.

Hytch also ensures your drive is carbon-neutral by investing in CO_2-reducing projects.[142] This is an additional benefit for riders and was attractive to their first partner, Nissan, which makes the Leaf electric vehicle. Nissan can put the CO_2 reductions earned from Hytch toward their corporate social responsibility goals.

Facilitating carpools isn't a new concept. In San Francisco, people create "casual carpools" by waiting at designated locations for drivers who want to use the HOV lanes. In Oregon, the Climate Trust launched a web page in the early 2000s to help people carpool and reduce carbon emissions. After five years, however, it reached only 4 percent of its goal.[143] These initiatives didn't achieve their potential, or failed entirely, because they couldn't connect people as easily and efficiently as smartphones can.

Smartphones have turned carpools from a coordination challenge into an opportunity to get paid. As Cleveland puts it, "Ultimately, Hytch offers an organizing principle for the idea that sharing a ride is valuable. The technology simply makes it obvious and easy to demonstrate and track." In short, it makes the idea a reality.

By adjusting rewards and tracking changes in ridership, Hytch can also find the sweet spot to reduce the greatest number of vehicle trips (and reduce CO_2 emissions) for the lowest cost. The US Department of Transportation study found that "a reward level of two cents per mile appears to yield indistinguishable results in monthly Trips per User from higher reward levels, but substantially better results than for lower reward levels." This is important to Hytch's corporate clients and helps governments looking to support the most effective traffic-reduction policies spend resources wisely. The study authors concluded that "monthly average awards of $7.54 for participants receiving two cents per mile . . . appear to be a very affordable cost as compared to other transportation investment options."

When we step back to consider these new developments, the incredible pace of the change becomes apparent. In 2005, mass transit looked very much like what was available in 1950. With smartphones, the options have exploded. Rahul Kumar, whose company models transportation patterns for cities, said, "I have never seen the market move as quickly as it has toward this demand-driven transportation mode. I think part of it is that people have an expectation now that things should be on-demand."[144]

Technology can also help navigate the growing number of transportation choices. In Vancouver, British Columbia, there are about 250 transportation providers. Some of them, like the Aquabus that carries people to Granville Island, are not always included on mapping apps. If you are trying to get around the city, it would be impossible to consider all the options—your car, the bus, a taxi, ride sharing, bike sharing, and so on. An app called Cowlines, named after the term transportation planners use to describe the best route somewhere, can help you decide. Users tell the app where they would like to go, and the app gives them options based on price, time, even environmental factors. Founder David Oliver recognized that people weren't able to take advantage of all the transportation options. In Vancouver "there are sixty-five companies designed to help people get from one place to another," he told me. "We asked people how many companies they thought existed and they would say only twelve." The app integrates all sixty-five of those options. It may tell you that the fastest route is to walk a few blocks, then take a bus, then use Uber for the final mile. Or use a bike-sharing program. The app allows you to choose the option that suits you best.

In Cowlines's beta test, they found dramatic time savings for users and reductions in CO_2 emissions. "We can beat Google Maps on a regular basis," Oliver claims. And because the app estimates the carbon emissions associated with each form of transit, it can track the savings user by user. During the test, Cowlines users reduced their emissions per trip by 52 percent. Ultimately, Oliver says the company expects the number to be in the range of 25 to 30 percent reduction overall. "That is huge," he said. "That would be two tons of CO_2 emissions per person, per year." And users don't have to give up time or convenience to achieve these environmental goals.

Because the CO_2 savings can be tracked, governments can use the data to meet their own carbon reduction targets, another key benefit. Whether political jurisdictions are using the Paris Climate Accord or local CO_2 targets as goals, their leaders are looking for ways to cut emissions. For Oliver, who previously worked to develop smart buildings, this is an exciting part of the app. He estimates that the cost of reducing one ton of CO_2 emissions using the app is currently four dollars. In the future "it will be about one dollar per ton," he predicted. This is far lower than existing approaches like subsidizing rooftop solar panels or biodiesel, which can cost over a hundred dollars to reduce a metric ton of CO_2.[145] Cowlines could dramatically reduce the cost to meet greenhouse gas targets.

In 2019, Greenlines, the developer of the Cowlines app, completed the first sale of carbon offsets generated by an app.[146] The CO_2 reduction results are so robust that in 2020, Verra, which created the most widely used system to verify emissions reductions, approved the methodology used by Greenlines to quantify the app's environmental benefits.

Cowlines isn't the only app providing green-travel options. Using research from the US National Renewable Energy Laboratory (NREL), Google also decided to offer users of their mapping app routes that save energy and reduce CO_2 emissions. NREL developed a model called RouteE to predict energy use over a particular road segment. The algorithm takes into account "anticipated traffic congestion, traffic speed, road type (including number of lanes), road grade, and turns."[147] Jeff Gonder, manager of the Mobility, Behavior, and Advanced Powertrains group at NREL, says the model is built on a database of driving behavior and models that estimate energy use for different powertrains.[148] The type of vehicle makes a difference. For gas-powered vehicles, keeping a steady pace is better than stop-and-go traffic. For hybrid or electric vehicles, stop and go traffic is not a problem since the regenerative braking systems help recharge the battery when drivers slow down. The system calculates both the rate of energy use as well as the overall usage. A long route at an efficient speed might still use more energy than a short trip in stop-and-go traffic.

In tests, Gonder found that about a third of trips had an opportunity for energy savings compared to the standard route drivers would choose. The fuel savings over that subset of trips amounted to 10 to 12 percent depending on the scenario. And although occasionally there was a trade-off between time and energy savings, Gonder said they found that in most cases there was "a double benefit of fuel savings and time savings."

Using that data, Google Maps changed the default to the route with the lowest CO_2 emissions, as long as the travel time is similar.[149] The model estimates it will "accurately select the route that consumes the least fuel 90 percent of the time."[150] And, because it estimates energy use, the model can also help electric vehicles improve battery range. The algorithm gives the option of choosing the fastest route, and—like Cowlines—provides other options, like biking or public transit, showing the mode that is most environmentally friendly.

Cowlines and the upgrade to Google Maps are good examples of how smartphones can aggregate a diversity of options, providing individuals with the ability to choose the best one for their particular needs while benefiting the environment.

THE EVOLUTION OF TRANSIT

With so much rapid change, transit agencies are recognizing they can also take advantage of the new technology. Transit managers have always understood that their transit options weren't suited for everyone, but it was the best they could do within budget constraints.

In some communities, like Altamonte Springs, Florida, transit agencies subsidize Uber trips.[151] "It is infinitely cheaper than the alternatives," like building roads or expanding mass transit, says city manager Frank Martz. Other cities include ride-sharing options in their transit apps. Brooks Rainwater, research director at the National League of Cities, says "it's exciting because ride-hailing can really serve this last-mile-first-mile challenge that so many cities have."[152] While mass transit may be good at covering most of the distance commuters need to travel, getting to a bus or rail station and then getting to your destination once you leave the bus or rail station (i.e., the first and last mile) is not included. Ride sharing can help.

Some cities are creating a hybrid transit–ride-sharing system. Microtransit options might include small vans to pick up people at locations throughout a city, similar to a bus but on-demand like Uber. The web page for Los Angeles's Metro system notes, "Unlike a standard bus, the service will follow turn-by-turn instructions from a navigation system that uses live traffic conditions and real-time requests for pick-ups and drop-offs to generate the most efficient possible shared trips for Metro customers."[153] The goal is to combine options for low-income residents with innovative approaches being developed by entrepreneurs. LA Metro's chief innovation officer, Joshua Schank, told *Forbes* their goals are to "make sure this service is accessible to everybody, that it's affordable, that it meets our standards for wages."[154] As microtransit pilot projects begin to launch across the country, each city will try something different. The diversity of solutions will serve different customers and allows experimentation to find out what works and what doesn't.

Microtransit isn't the only innovation made possible by the ability to rapidly share information using smartphones. A company called Pantonium is offering what they call "on-demand macro-transit" as an alternative to microtransit. Their system promises to pick you up near your home or work and take you to your destination quickly and without unnecessary stops or transfers, guaranteeing you arrive at the time you indicate. It is a win-win-win: it is more responsive to rider demand, costs taxpayers less, and reduces environmental impact.

Remi Desa, Pantonium's CEO, wanted to change the way transit worked, updating it from an approach that had been in place for decades. "When you look at how public transit was designed, it had hub-and-spoke where parts of the city were connected to the center of the city," Desa told me. Most transit riders would take one bus to a central location, then transfer to another bus that would take them to their destination. This increases the time it takes to get somewhere and requires more buses to cover all parts of the city. It was a trade-off between providing enough geographic coverage so riders could find a bus stop near them and keeping costs down by serving only locations where there is enough demand.

Pantonium decided to create an on-demand system, creating bus routes on the fly using an algorithm. Passengers download the app from a community transit agency, provide contact information, and set up payment. When people need to take a trip, rather than going to a bus stop and waiting, they book their trip through the app, receiving an estimated pickup time and location that should not be more than three blocks away. They can track their bus while they wait. Once the bus arrives, passengers pay the fare using the app. "You put the information in about where you are going and when you need to be at work," Desa explained. "Our system looks at the capacities on the buses, where everyone is going, what time you need to get to where you need to get, and we are calculating all these permutations."

Like a normal bus, it may pick up riders at other stops along the way. On-demand transit, however, will pass bus stops where no users have requested to be picked up or dropped off. Pantonium estimates that between 60 and 70 percent of stops can be bypassed in a typical run, saving time and money. Transit agencies use only the buses they need, preventing buses from burning fuel on unproductive routes. And, because it minimizes stops and transfers and increases the service area, it attracts new riders because the service is faster, more convenient, and more reliable.

On-demand macrotransit differs from microtransit because a transit agency uses its existing full-size buses but optimizes technology to provide better service. Using full-size buses, Pantonium argues, allows for greater ridership and efficiency.

The community of Belleville, Ontario, was the first to use the system, testing it with evening service. It was a perfect test for the flexibility of the system. "In the evening you have shift work or people going to different places," Desa explained. "It is really difficult if you don't have an understanding of where people want to go." Belleville's system was a traditional hub-and-spoke system, where riders would have to ride to the center of the system and then transfer. As described by a Belleville Transit official, "[Riders] no longer have to plan around us. We get to plan around them.

They have the power and ability to book the bus when they need it, where they need it, to go where they need to go. That alone… is of huge value."[155]

Once users began to provide information about the routes they needed to travel, the city realized its existing bus routes didn't match what residents wanted. "What we found is that first of all the city didn't really understand where people wanted to go," said Desa. "Very fast they increased their ridership about 300 percent." The service area also expanded by about 70 percent and, remarkably, the total fleet mileage decreased by 30 percent.

The community of Stratford, Ontario, saw an even more dramatic improvement. Their weekend-only pilot saw the distance traveled by their buses on Saturdays fall by two-thirds, from 2,200 kilometers down to just 720, saving more than $600 each Saturday.[156] The reduction in fuel use represents not only savings to the taxpayers but also less CO_2 into the atmosphere.

After the success in Belleville, the city of Sioux Falls, South Dakota, decided to see if the system could improve ridership and save money. Allie Harzler of Sioux Area Metro explained that three-quarters of their riders had a smartphone and that the city needed to find ways to economize. "We are facing a budget deficit if we don't figure out a more sustainable solution," she explained. "The hypothesis is that this will help us figure out if we need full-size buses or can we buy smaller buses which are more nimble and better for the environment." Soon after they implemented the project, the results were similar to those in Belleville. After launching a pilot project on the weekends, the number of buses needed fell from nine to six, reducing the average trip time, and completely eliminated passenger transfers.[157]

The combination of increased ridership and fewer miles traveled also reduced CO_2 emissions. The increased accessibility encouraged people to take transit rather than driving alone to work or to the store. By using fewer buses and driving fewer miles, the city also used less fuel. Pantonium estimates that by 2022, this system could "help prevent over 6,000,000 car trips," saving more than 20,000 metric tons of CO_2, equivalent to the emissions of about 5,000 cars a year.[158]

The flexibility of the system makes it attractive to riders and makes on-demand transit more durable as a long-term solution. "Cities are constantly changing," noted Desa, "and when you see new employers come into town, it is hard for transit agencies to adjust their routes. You see these routes that start off being productive and end up becoming unproductive as they add stops to meet new demand." He notes that when a new employer or a new store—"Walmarts are always the most popular stop we go to in any city"—comes into a community, it can change mobility patterns. Those new patterns will become clear as users request different trips, allowing the transit agency to adjust quickly.

Responding to consumer demand also helps ensure that the environmental benefits of the system are immune to political changes. Traditionally, cities have promoted public transit by making alternatives more costly or difficult. For example, Seattle proposed making a "major area" of the city "restricted to most cars."[159] Other places have imposed carbon taxes, congestion prices, or reduced parking spaces to discourage driving. These policies may endure in some places, but they are politically fragile and have been repealed in more than one community. For example, a fuel tax designed to cut France's greenhouse gas emissions sparked the "yellow vest" movement in France that led to riots and caused President Macron to back down from the proposal.[160] Offering a low-cost, environmentally friendly alternative to driving, on the other hand, isn't dependent on the results at the ballot box. The most ardent skeptic of climate change doesn't have to be convinced to keep a system that increases ridership and reduces costs.

Reducing exposure to political change is an important reason these new transportation options are likely to endure. Diversity is durability.

Calestous Juma, a scientist and expert in sustainable development who served as director of the Science, Technology, and Globalization Project at Harvard's Kennedy School of Government, argued that innovation that engages consumers and offers versatility is key to promoting sustainable policy. "The essential message for policymakers," he argued, "is to explore ways by which they can choose technological options that offer greater versatility. This can be achieved through consumer preferences." In their own way, car sharing, microtransit, improved mapping apps, and macrotransit respond to consumer demands. Although none of them individually will solve the challenge of transportation-related CO_2 emissions, in combination they can be adapted to meet the needs of most people, rewarding them for being environmentally conscious.

REDUCING THE TRADE-OFF BETWEEN ENVIRONMENTAL CONSERVATION AND PROSPERITY

Saving money also helps facilitate environmental choices by reducing the burden of competing financial demands.

By creating a variety of choices that fit many lifestyles, innovative environmental apps reduce the trade-offs that can inhibit environmental choice. Research demonstrates that when people feel there is a trade-off between a preferred option and the environment, they will avoid unpleasant information and make the choice that suits them, not the environment. For a 2006 study on willful ignorance, Kristine Ehrich and Julie Irwin asked participants to choose between products, selectively providing them with information about the price, desirability, and environmental impact of each to see what information the participants prioritized.[161] In one example, they asked participants to rate sixteen theoretical desks, comparing them on workmanship, comfort, ethically sourced wood, and price. Subjects were required to ask for attribute information desk by desk but were given only limited time, so they were instructed to ask about the attributes most important to them. For the ethically sourced wood category, desks could be built using wood from tree farms that did not impact the rainforest, or a mix of wood from tree farms and the rainforest, or wood that depleted the rainforest. After ranking the desks, participants were asked one final question:

Which of the following best represents your view about rainforest wood?

1. Use of rainforest wood should be determined by the free market.

2. Protection of rainforests should be balanced against market values.

3. Protection of the rainforest should be absolute.

Ehrich and Irwin found that the more important ethically sourced wood was, the *less* likely a participant was to ask about the ethical attribute that protected wood from the rainforest. They noted, "Participants who stated that they care very much about the protection of rainforest wood were the most likely to show a large discrepancy between their request for and use of the ethical attribute information." People who said that protection of the rainforest should be "absolute"—that it should override all other considerations—were only slightly more likely to ask about that attribute than others who said it made no difference. Participants avoided information that could make their decision difficult. It was deemed easier to buy a low-cost, well-made desk if you didn't know the wood source than if you knew it was harmful to the rainforest. Recognizing that they would have only limited information about their options, those with the greatest emotional stake in environmental information avoided the question. Environmental ignorance was bliss.

Having complete or near-complete information about the range of options helps overcome that problem, allowing the selection of a choice that most closely matches a consumer's values and needs. If participants had access to all information about the desks, they could have quickly selected the option that best met their goals for cost, quality, and environmental impact. This is what the Cowlines app does for people looking to get around Vancouver. It quickly aggregates all information, simplifying the wide diversity of options, so people can make the best choice. It doesn't completely eliminate the trade-off between cost, time, and environmental benefit, but it narrows the gap.

This is why agency—the power to act—that Scott Page emphasizes is so important. By providing options to address potential concerns, people become less afraid of the information. Page notes, "People don't have an awareness of what their own effects are. By getting people involved in these discussions, they get an understanding of the algebra of environmental impacts." He encourages us to "be involved in the decision." Smartphones make that more possible.

6

How Transparency and Blockchain Make Fish and Chocolate Better (But Not Together)

Guido van Staveren doesn't mince words. "It is not about hope. Cut the bullshit. No storytelling. Show me the proof." As founder of the FairChain Foundation, an organization promoting environmentally sustainable business practices, van Staveren believes transparent supply chains are key to fighting deforestation and reducing CO_2 emissions. To test that theory, he created companies that sell chocolate and coffee. Being an environmentalist doesn't mean you can't eat well.

It seems strange to say that something as mundane as transparent accounting can be a powerful tool to address big problems like deforestation and climate change, but as consumers and businesses look for environmentally responsible products, they want to know they are getting what they paid for.

This fight is being played out at the highest levels of climate policy. In crafting CO_2 reduction regulations, there is a heated debate about the role "carbon offsets" should play in reducing emissions. Carbon offsets are simply projects that reduce CO_2 emissions compared to a business-as-usual baseline. Rather than allowing landfill methane (a greenhouse gas that is up to thirty-six times as potent as CO_2[162]) to go into the atmosphere, there are projects that capture it and use it as an energy source, reducing the need for other fossil fuels. Those who want to reduce their own carbon footprint or must reduce emissions to meet regulatory requirements can invest in carbon offset projects. Since a ton of atmospheric CO_2 has the same global warming impact whether it is emitted in Tuscaloosa or Timbuktu, projects that reduce emissions anywhere in the world can counteract the CO_2 I emit by heating my house or driving to work. This seems strange to some, but it is no different than paying someone else to grow food rather than doing it yourself. If someone else can do a better job of reducing atmospheric greenhouse gases, we should pay them to do it rather than try to do it badly ourselves.

There is a catch. While I can hold a tomato someone else has grown in my hand, it is difficult to know with certainty that a project claiming it reduces CO_2 actually does what it says. Credible and transparent accounting is crucial. This is why carbon offsets are attacked from the left and right.

From the left, environmental activists argue that phony carbon offset programs allow companies to continue to emit CO_2 emissions without actually reducing total worldwide greenhouse gases. For example, in California, forests set aside to absorb CO_2 were lost to forest fires, with much of that stored carbon being re-emitted into the atmosphere.[163] Meanwhile, the emissions cap in the state's climate law assumed that CO_2 would be safely stored for decades. The loss of those forestry offset projects may mean the state's total CO_2 emissions are higher than the law would otherwise allow.

From the right, offsets are characterized as "indulgences," much like the religious salvation sold by the Catholic Church in the sixteenth century. The insinuation is that purchasing offsets is cheating and isn't a legitimate way to actually reduce the environmental impact of those who believe we face a climate crisis.

While both critiques have their ideological element, each is rooted in the belief that offset accounting is sketchy. Faced with a lack of transparency, ideology fills the vacuum. In my home state of Washington, environmental activists who oppose the use of offsets have also been strong advocates of government subsidies to increase the amount of wind energy and state programs to pay farmers to capture agricultural methane. Ironically, both of those types of projects are available as privately created carbon offset projects.[164] What is the difference between a project that captures agricultural methane funded by government and one that is funded by private investors? Nothing. Without transparent accounting, however, people turn to other ways to assess credibility. How you feel about government or markets influences how you judge otherwise identical projects when other evidence is opaque.

Technology can help create the transparency necessary to move from faith to data. I buy carbon offsets that capture methane from landfills because I think they are more reliable than forestry offsets. The math is relatively straightforward and tangible. I would need to see some pretty robust accounting to buy offsets from projects where the emission reduction calculation is more complex. If I don't believe the accounting, I won't buy it. As van Staveren says, "Show me the proof." Technologies, like supply chain transparency using blockchain accounting, can make that proof widely available to environmentally conscious consumers.

The key, however, is not the particular technology. "Why blockchain," van Staveren asks rhetorically. "We don't care about technology. We care about the problem." To him blockchain technology "is a lie detector. It delivers something that could lead to trust." Earning the trust of consumers opens a vast range of opportunities for sustainable businesses to fight deforestation, overfishing, and climate change.

BLOCKCHAIN FOR CHICKENS?

The word "blockchain" is thrown around a great deal, but not everyone understands what it is. Most associated with cryptocurrencies like Bitcoin, blockchain is a technique of decentralizing information storage so it is transparent and independently verifiable. That transparency makes it possible to ensure the information is reliable. It is being applied in many new ways.

Before ordering chicken at a restaurant in the TV series *Portlandia,* based on the quirky, ultra-environmentally conscious culture in Portland, Oregon, two customers have questions: "Is that USDA Organic, Oregon organic, or Portland organic?" The waitress brings them the chicken's "papers" to reassure the customers that the chicken they will be enjoying was raised locally, was fed organic hazelnuts, and had four acres on which to happily roam. And that his name was Colin.

I live in the Pacific Northwest and can confirm that this kind of exchange is not much of an exaggeration. Nor is it limited to Portland. People want their chicken fed an organic diet. They want to know if it was injected with antibiotics. They want to know if it was free-range and, if so, how large the range was.

It is now possible to make that information available instantaneously. "In China, now consumers who want to feel good about their chickens have another option: free-range and organic birds raised with an anklet that tracks and reports every aspect of their lives," and all that information, they claim, will be available to consumers using blockchain technology.[165]

Having certainty that your chicken, or tuna, meets verifiable high standards can be a challenge. In an interview with *Science* magazine, Guillaume Chapron, an ecologist at the Swedish University of Agricultural Sciences, explained how uncertainty in verification systems makes it difficult to ensure environmental protection.[166] "If you buy a fish at the supermarket, the supply chain is very long. The supermarket might not even know where it came from," he explains. "And so, there are multiple opportunities for environmentally unsustainable goods to enter the supply chain." Using blockchain, information about a fish can be collected at each step of the process so consumers can be sure it came from where the seller claims it does. "A blockchain-based supply chain would mean that when you buy a fish, you scan a QR code [like a bar code] with your smartphone, and you see every step," says Chapron. "And you know that it cannot be falsified."

Currently, people turn primarily to the government to ensure these kinds of standards are met. The US Department of Agriculture created the USDA Organic standard in 2002 to formalize assurance that products claiming to be organic were subject to inspection and companies would be held accountable if they weren't actually meeting those requirements. An independent certifier using blockchain technology can now provide that certainty and combine it with the diversity of choices desired by consumers.

Rather than creating a one-size-fits-all organic label, these new options may allow people to pick and choose the attributes they care about. Rather than USDA Organic or Oregon organic, you can choose your own version of organic.

The range of options on the horizon is incredible. The CEO of ZhongAn Tech, the company behind the effort to track chickens in China, told the *South China Morning Post*, "Each of our chickens wears an anklet since birth, which is an IoT device that connects wirelessly to our blockchain-based network and sends real-time data about the bird's whereabouts, and how much exercise it gets every day."[167] That's not all. The level of detail available makes the *Portlandia* sketch seem tame by comparison. One report on the project noted, "The chicken's age and location, how far it walks each day, air pollution, the quality of the water it drinks, when it's quarantined, when it's slaughtered, and other details, are all recorded in the blockchain." Want a chicken that enjoyed only clean air, ate organic feed, and jogged a mile a day? Soon, you may be able to select for that. Three years later, the system was still not commercially available, so the technology is more about possibility than reality. But it isn't the only example.

In the United Kingdom a company called Provenance developed an eponymously named smartphone app that's blockchain-enabled to help consumers know where their food is coming from. Rather than rely on a single certification source, like the USDA or an independent nonprofit, using blockchain allows Provenance to ensure transparency and security across the many transactions in the food production process. The company says in a white paper, "In the face of these efforts, we must ask ourselves: can one organization be trusted to broker all data about every product's supply chain? The truth is that no single organization can, and that relying on one party (or even a small collection of cooperating parties) creates an inherent bias and weakness in the system."[168] The danger is that centralized systems create the opportunity for corruption and make it difficult to audit independently. Centralized systems also create a single target for hacking. Furthermore, it is difficult for one system to be at every point in a supply chain to ensure nothing slips through the cracks.

Provenance believes a blockchain-based system can address those problems. In a case study, they tracked sustainably harvested tuna from Indonesia to the consumer, testing the ability of a blockchain system to follow every step of the process. From the time the tuna was caught until it was transferred to the processor, a record was added to the blockchain. The Provenance team noted, "Every fisherman, supplier and factory worker we met had a mobile phone."[169] A local NGO can audit the participants, using the transparency of the blockchain to see who caught the fish and how it traveled. None of the individual participants control the process, making it more transparent and robust.

And it makes tracking more flexible, providing the opportunity to cater to customers. The market for ethical shoppers is growing and consumers want assurances they are getting what they paid for. Provenance notes that a transparent blockchain would allow tracking that "extends to dining environments, indicating available information on ingredients." Rather than handing papers to a diner or shopper, a waiter or grocery service person could simply show them the blockchain evidence.

Appealing to selective customers is why the Grass Roots Farmers' Cooperative in Arkansas partnered with Provenance to track the shipment of their chicken. "Until now, it's been a struggle for us to tell the story of why our foods are different from those raised in feedlots and large chicken houses," Grass Roots explained on their web page.[170] "With blockchain, we can show you." Scanning a QR code with the shipment, a customer can watch as the chicken travels from one of Grass Roots' partner farms, to a processor, the co-op, and then finally to the San Francisco buyer. There's even a picture of the chicken.

The transparency provided by blockchain-based systems could also help prevent illegal fishing. In 2020, a large fleet of Chinese fishing vessels approached an exclusion zone around the Galápagos Islands, looking to fish in the waters for a range of species, including the endangered hammerhead shark. Although Ecuador stepped up enforcement of their waters, preventing illegal fishing can be difficult and Chinese boats have been caught in the exclusion zone in the past.[171] Transparency can help consumers choose fish that have been sustainably caught, avoiding fish with a murky origin.

Kenneth and Shaunalee Katafono launched TraSeable, a Fiji-based organization that uses blockchain to track fish from "bait to plate," sharing the information with buyers and consumers so they know exactly how and where the fish they are buying were caught. They partnered with WWF to test technology that would improve the transparency of fisheries operations. When a fish is caught, it is immediately tagged with an RFID tracker and scanned into a blockchain-based database. As each fish moves from catch, to process, to market, more information is added to its data chain.

Kenneth Katafono told me, "When the fish trades hands, data about change of ownership is reported in the blockchain. This is fairly standard in the food industry." His focus is on developing countries and small operators, where "traceability is uncommon." One fish can move around the world and regulations can be different at each step of the process.

Providing a reliable record allows consumers to know what they are getting even if it passes through countries where regulation or enforcement is poor. "The blockchain just provides that layer of trust that all the data [are accurate], regardless of the actors along the supply chain back to where the fish was caught," said Katafono.

Rather than simply relying on the honesty of fishers or others, technology can be used to ensure the quality of the information. GPS data showing a boat's location can be added to the blockchain record so captains know if they move into illegal fishing grounds, there will be a record.

In the case of the Chinese fishing fleet, it is unlikely it would submit to this kind of scrutiny and there is not much of a market for hammerhead shark in the United States or Europe, where consumers have the disposable income to be selective about what they purchase. Katafono admits this is an obstacle. "A lot of the fishers we see implementing traceability—they are the good guys. For them it is an additional way to prove they are doing something good," he explained. "The challenge is getting [to] the ones who aren't doing good."

There is precedent for reducing the environmental impact of tuna fishing by using information that highlights the good actors. The "dolphin-safe" label on tuna cans became the standard for all brands. Any canned tuna without the label is simply unlikely to be purchased by consumers. Even if bad actors don't provide their information, consumers could simply avoid products without evidence that a fish was caught legally, just as nobody wants to buy a can of tuna that isn't certified dolphin-safe.

The new technology provides an additional level of certainty. "Ecolabels are good and serve a purpose," says Katafono, "but if you understand how ecolabels work and the lack of monitoring on the ground, there are gaps." Technology that increases transparency can fill those gaps, helping ecolabels live up to their promise. Ultimately it is up to selective consumers to choose products they know came from reliable sources.

USING TRANSPARENCY TO FIGHT DEFORESTATION AND CLIMATE CHANGE

Something similar is being done for chocolate bars. Working with the United Nations Development Programme (UNDP) and the FairChain Foundation, farmers in Ecuador created "The

Other Bar," which uses blockchain to connect the consumer and farmer. Each chocolate bar includes a scannable token that gives the buyer the choice of a small refund that can be used toward more chocolate or putting the refund toward buying a tree in Ecuador. Farmers also receive a larger percentage of the income from each sale.

The combination of increasing the revenue to farmers and planting trees helps the environment in two ways. First, the revenue reduces the pressure to cut down the native forest to expand their plantation. Second, planting trees increases the amount of CO_2 that gets absorbed.[172] If you choose the tree (as 85 percent of buyers did), FairChain tracks the purchase using a blockchain system to show where your tree is, from purchase to delivery to planting. FairChain also provides buyers with a personal "impact monitor" that calculates how much CO_2 will be reduced by your tree and the trees purchased by others.[173]

There is a strong connection between deforestation and CO_2 emissions, and reducing the amount of land lost to deforestation has several environmental benefits. At the beginning of the UN Climate Change Conference in 2021, more than 130 nations signed a declaration on "Forests and Land Use" that highlighted the value of forests to "help achieve a balance between anthropogenic greenhouse gas emissions and removal by sinks; to adapt to climate change; and to maintain other ecosystem services."[174] A primary driver of deforestation is poverty, as trees are harvested to cook food or heat homes, or are removed to make way for agriculture. Indeed, the announcement recognized the need to address these economic pressures, saying the countries acknowledge they must "enable sustainable agriculture, sustainable forest management, forest conservation and restoration, and support for Indigenous Peoples and local communities." Projects like The Other Bar are designed to make forests economically profitable for farmers, reducing the pressure to convert the land to other, more lucrative—but environmentally destructive—purposes.

I purchased five of the bars to see how the system worked. I believe one cannot fully understand the system without testing the quality of the chocolate. It is the kind of sacrifice I am willing to make for the environment.

After I scanned the token included with each bar, the system displayed a map of the village in Ecuador where my tree would be planted. For every four tokens, one tree is donated to a farmer, creating nineteen pounds sterling in revenue for a coffee farmer and a tree that will absorb CO_2 during its life span.

In addition to helping Ecuadorian cocoa farmers, The Other Bar tested the concept that consumers will choose a particular product in order to feel a little closer to the people who produce their food. Guido van Staveren founded the FairChain Foundation, which partnered with the United Nations Development Programme (UNDP) to create The Other Bar. "The Other Bar was an experiment," says van Staveren. The experiment was a success. "We sold 40,000 bars in a couple months. It shows the mechanism works."[175] The blockchain gives consumers information about the sustainable choices they are making and provides incentives to keep making those choices. It provides consumers with information about where a product comes from while giving them the opportunity to redeem tokens for rewards or to promote environmental benefit. It makes engagement good for the planet and good for the company. "You can turn marketing into tokens," says van Staveren. "Give tokens to consumers and empower them to be part of the impact goals of the company. Hopefully, that will lead to happier customers." The lessons from The Other Bar were put to work when van Staveren started Moyee Coffee. By roasting the coffee in Ethiopia, where the beans are grown, a greater portion of the value of a cup of coffee stays in that country. Moyee Coffee customers can also use the unique token they are given to plant new coffee trees, reducing CO_2 emissions and helping farmers. "The trees we plant and forests we plant are climate positive," he explains. "It absorbs more CO_2 than it uses."

The tokens make clear the symbiotic relationship between sustainable practices and sustainable business. "We only plant trees that earn cash," says van Staveren. Rather than planting trees in a forest preserve that may be cut down later, the trees Moyee Coffee planted using the tokens create economic value. "The only way to protect forests is to make sure they have earnings," he explains. When consumers use their token to plant trees, they also become fans of the brand, creating an opportunity for additional marketing and sales. "About 15 percent of people plant a tree and leave their email address," he says. "This is a huge result for a marketing campaign."

That model is possible because the technology is now available to small businesses. "There is a growing group of people in companies who are purpose-driven," van Staveren notes. "We are not alone. Doing that is hard work and this is done mostly by small-scale companies who really care." Small technology made that possible. "We have a technology suite that is low-cost. A huge number of small companies can start in a very easy way."

Projects like these demonstrate that blockchain can be used to empower both producers and consumers while improving the market for products that are environmentally friendly.

Information transparency is important for another reason: as with transit options, it can provide a more honest accounting of environmental impacts. For example, it could reveal that locally grown food used more land, water, and resources than food grown where the climate and soils are best. Research by Dr. Steve Sexton of Duke University found it would take a farmer in Michigan fourteen acres to grow the strawberries a California grower produces on a single acre.[176] And it would take an additional sixty million acres of land, an area the size of the state of Oregon, to grow forty crops entirely locally in the United States. This is surprising to many because it would seem to make sense that the shorter distance food is shipped, the smaller the environmental impact. Transportation, however, generally accounts for less than 10 percent of energy associated with produce. Growing food where it makes sense—strawberries in California, pineapples in Hawaii, potatoes in Idaho—more than makes up for the small additional energy to transport the food to its destination. Blockchain-based tracking can help reveal this kind of counterintuitive information.

Environmental impact isn't the only factor people consider when buying organic produce or produce from a local farmer. Some like the taste (raw honey from local bees tastes better than mass-produced honey from the store), want to buy from someone they know, or want free-range chickens rather than ones that live their lives confined in coops. With information transparency, people can weigh these factors for themselves. As the diversity of options increases, transparency makes it possible to select the best option from a range of environmental choices. Transparency makes diversity functional.

My hope is that in addition to tracking how many acres a chicken roamed, blockchain technology will allow consumers to see how much water and fertilizer it took to grow the crops they are purchasing. Currently, people look for certified sustainable labels because that's the best system we have. These labels are often misleading or incomplete. It is difficult to tell a complex story with a label that simply says, for example, "organic." USDA Organic allows pesticides. Other organic systems do not. Not using pesticides may mean that more of the crop is lost, along with the water and energy that was used to grow something that ended up as bug food. A simple label does not convey these nuances and trade-offs.

As technology makes it easier to track and verify all of the attributes of food, the more information consumers have available, the more likely they are to make decisions that effectively help reduce damage to the environment.

THE EXPONENTIAL POWER OF MICROCONTRIBUTIONS

Individual acts and decisions—like choosing sustainably caught tuna—can seem inconsequential. Our ability to collaborate, however, acts as a force multiplier, making each small action more powerful. One ride in a Zipcar may seem to do little, but in the aggregate, ride sharing has significantly reduced resource use and CO_2 emissions. But the benefits of those actions add up only if the contribution each makes to the environment is real. With small efforts, the margin for error is also small, so it requires increased certainty that the benefit to the climate, to fisheries, or to forests can be proven. With increased transparency provided by the blockchain and other technology-based systems, the evidence is available and reliable.

In his book *Reinventing Discovery*, Michael Nielsen argues that technology lowers the cost of contributing to solutions.[177] Microcontributions, as he calls them, add up over time. "Microcontribution lowers the barrier to contribution, encouraging more people to become involved, and also increasing the range of ideas contributed by a particular person." His focus is on the ability of individuals to contribute ideas and expertise, but the same is true of actions that promote environmental improvement.

People rarely consider taking small steps because it can take a good deal of time and effort to acquire information, weigh the options, and make a decision. The more effort required, the greater the impact we want to see as a result. By way of contrast, if the burden of finding a solution that fits you is low, you will be more willing to take many small steps that add up to large environmental benefits.

As in the standard physics equation, the power of your impact is equal to the force times velocity. Even if the force of each decision is small, the frequency (i.e., velocity) at which you and others in the community take those actions makes your efforts more powerful. Smartphone technology that provides a diversity of informed options, the power to act immediately, and the knowledge that each of those actions is making a real difference, increases the velocity of your actions, making each one more forceful.

What the Flint Water Crisis and Thomas Edison Have in Common

Here is your assignment: Buy pizzas for a hundred people, ensuring there are slices with toppings that satisfy everyone and everyone eats as many pieces as they would like. Difficult, but not impossible. You have one minute to complete the task. Now it is impossible. The difficulty of what is a fairly straightforward task says a great deal about why our electrical system is structured the way it is and why governments have traditionally acted as surrogates for individuals when addressing environmental problems.

Smartphones and information technology are changing that.

Not far from where Thomas Edison built the nation's first power plant, homeowners are generating energy and selling it to their neighbors, much as Edison's Pearl Street Station did a century and a half ago. In Brooklyn, homeowners with solar panels are finding they can use the electricity they generate or they can sell it, earning a bit of extra cash, sharing surplus solar energy in the middle of the day when they do not need all that is being produced. Much like Edison's first small grid—and very much unlike the massive grid most people are connected to—the Brooklyn Microgrid connects local energy producers with people who need it. In addition to providing renewable energy, the managers of the Brooklyn Microgrid say there is appeal in buying from people who live nearby.

Scott Kessler served as director of business development at LO3 Energy, the company that created the Brooklyn Microgrid. He explained that buying energy from a neighbor's solar panel is about more than just the environment. "You can go buy environmental credits from a number of retailers, at least in a deregulated state like New York," he explained. "Here we can point to where those credits are going, at least in the first stage. The local aspect is what makes it different." Like coordinating the pizza choices of many people, coordinating energy transfers, even across a small area, sounds simple, but the cost of gathering the information and making quick decisions about when to buy and sell electricity can be very high, making it functionally impossible. Technological improvements have made that difficult task easier, allowing people with more energy than they need to sell it to others virtually immediately, while providing an incentive to conserve. The more you conserve, the less you have to buy and the more you may have to sell.

Microgrids aren't new. Edison's notion of having small, direct-current power stations connected to a limited number of businesses was a kind of microgrid. When alternating current allowed electricity to travel longer distances, utilities took advantage of the savings that came from increased scale and built large power plants to supply many homes and businesses. With consolidation came fragility. In her book about the system that delivers electricity in the United States titled *The Grid*, Gretchen Bakke writes that many of America's largest blackouts have been caused by small errors that had widespread impacts. She recounts as examples "the 1965 Northeast blackout, caused by an incorrectly set relay just outside Niagara"; the "2011 blackout in the American Southwest which propagated from a lineman's error in Arizona"; and "a 2014 blackout in Detroit—home of the nation's least reliable municipal grids—when a single cable failed.[178]

That type of fragility is what initially inspired the idea for the Brooklyn Microgrid. "We decided to start in Brooklyn because it is diverse and was hard hit by Hurricane Sandy," explained Kessler.[179] Generating some electricity locally would make communities less vulnerable to infrastructure problems outside their control. In the United States, microgrids are still connected to the larger grid and rely on that system to supply most of their electricity, especially at night when there is no solar power being generated. During a blackout, a microgrid can include solar panels, diesel generators, and other sources of energy that fill in until the power is restored.

Microgrids could be more than just a backup source of energy in an emergency. As the Brooklyn Microgrid and others across the globe are demonstrating, the technology makes it possible for people to act as little utilities. The *New York Times* explained, "The ability to complete secure transactions and create a business based on energy sharing would allow participants to bypass the electric company energy supply and ultimately build a microgrid with energy generation and storage components that could function on their own, even during broad power failures."[180] By creating a little market among neighbors, microgrids create options for both buyers and sellers. Buyers can now purchase when they need it or can select the attributes they are looking for, such as clean energy, and pay a bit more. This is possible because the technology can keep track of sales between customers in a way that is simple and automated. Whether using the blockchain or some other electronic ledger, a microgrid makes it possible to buy and sell electricity without constant oversight. "Individuals could say how much they are willing to pay, and we could say whether it is enough to buy some solar power, or whether they are buying from traditional sources," explained Kessler. "Maybe in the next fifteen-minute segment you will be able to afford it." Likewise, those who want to sell excess electricity can market it to be attractive to buyers. "For any person who is putting energy on the grid, we should be able to get a value for that energy, for time and location and where it was created—solar panel or diesel generator." The role each person plays can change. Those out of town on a hot day can sell the solar energy they aren't using to power someone else's air conditioner. During a blackout that same person may decide to buy energy from their neighbors on the microgrid so they can enjoy their morning cup of coffee.

For those interested only in environmental benefits, there are already options, known as renewable energy credits, or RECs. When someone buys an REC, they are buying a promise that that amount of electricity, typically a megawatt hour, has been generated and put on the grid by a renewable energy source, displacing a carbon-emitting source like coal or natural gas. In the same way you don't withdraw the actual dollars your employer puts in your bank account at an ATM, you probably won't use those specific, renewably generated electrons. Instead, you are guaranteeing that the necessary amount of renewable energy is on the grid, which changes the overall mix of electricity. Microgrids create opportunities that don't exist with the current regulatory system promoting renewables.

If you have solar panels on the roof of your home, some utilities are required to buy a portion of that electricity and sell it on the market in a system called "net metering." Frequently, homeowners receive a very high price when selling their electricity to the utility, one that is far above market rate. The difference between retail price and the rate paid under net metering is paid either by other ratepayers or taxpayers in the form of subsidies. That may sound fine to some, but there are two big drawbacks.

The first is fairness. Solar panels are expensive to install and are not generally available to renters. Subsidizing rooftop solar through net metering ends up charging working-class families so those wealthy enough to own solar panels get a little more money. In California, for example, energy researcher Ben Zycher from the American Enterprise Institute found that in 2016 net metering "subsidize[d] the affluent (median income of those installing solar systems: $91,210) at the expense of all other power consumers (median of $67,821)."[181] It is one thing to promote renewables. It is another to tax low-income families to subsidize the wealthy who could otherwise afford the purchase without subsidies.

Second, net metering treats all electricity, no matter when it was generated, the same and gives those selling energy a flat fee. Net metering can also put an annual cap on the amount sellers can receive as subsidies. The Brooklyn Microgrid uses a different approach, recognizing that the price of electricity is highly dependent on when it is available. "States that have had [net metering] are moving away from it. It is not a good way to value energy," explained Kessler. "There is a time and location component, and we need to value that." On the Brooklyn Microgrid, "it isn't going to be a flat fee. The price is going to reflect the impact you are having on the grid." In some cases, that may be less than the subsidized price available to sellers in net metering. At other times it may be higher. And since there isn't an annual cap, there is profit available to those who find ways to save electricity and offer the excess to others on the microgrid. That's exactly what happened during the Texas energy crisis in 2021, with homeowners saving energy so they could sell more of the energy from their rooftop solar panels when prices skyrocketed.[182]

Microgrids put customers at the center of the electricity equation. With net metering, even though individual homeowners can sell their excess energy, they have to sell it to the utility. The overall system is still the same, but with a small modification. Microgrids aren't going to replace that larger system in the near future—or maybe ever—but they give people options. The creators of the Brooklyn Microgrid are optimistic about the future. Kessler told me:

> You are seeing a shift from the utility being the focal point of everything and not being open to other ways of doing things. Now you are seeing a lot of openness and seeing others as

partners. The customer used to be called a 'ratepayer.' This weird legal word. Now they are looking at customers differently. We can get away from what was a purely vendor relationship and now I am providing a service that the customer really wants and likes.

The ability to rapidly share data is driving that change. Rather than the utility controlling information about supply and demand, customers and sellers can see the opportunities and let an automated system make decisions for them within guidelines they set. Although distributed energy generation, like rooftop solar panels, was important to creating the idea of microgrids, distributing the information about electricity supply and demand made them functional.

The simplicity of the technology allows innovators around the world to create microgrids. The company behind the Brooklyn Microgrid, LO3 Energy, also tested a microgrid in Australia.[183] The homes were connected to the grid to provide electricity when solar is not available.

In Estonia, another utility is also looking to improve energy markets using blockchain. Although the system isn't a microgrid, the concept is the same, with customers connecting directly to independent electricity providers and sharing data at a larger scale. Giving consumers more options is the goal, but it could also help increase the percentage of renewables used by the country. As one description of the project highlighted, "The bulk of energy in the Baltic country is produced by fossil fuels—only 18 percent come from renewables. The main aim is to test the limits of what's possible with blockchain technology."[184]

Decentralizing the system also makes it easier for companies to move quickly to renewable sources of energy without having to enter into long-term contracts or alter existing contracts. One report about Estonia's effort noted, "There is an incentive to increase the number of energy applications using blockchain:

it helps to decentralise the grid for more and more distributed energy resources, which are largely renewables." Fei Wang, a research analyst at the energy research firm Wood Mackenzie, explained, "Blockchain as a technology can provide an alternative way to manage the increasingly more decentralised system." Rather than contracting with one supplier and being subject to their contract restrictions, using distributed energy offers the flexibility for companies to choose the mix of energy sources, price, and reliability that suits them.

In Bangladesh, a company called SOLshare was given a grant by the United Nations to develop microgrids, providing electricity where there may not be any.[185] The microgrids in Bangladesh aren't that much different from Brooklyn's—a solar-powered microgrid that allows purchases and sales using a distributed ledger. Indeed, the description of the program could be from anywhere in the world: "Our decentralised peer-to-peer microgrids deliver solar power to households and businesses and enable them to trade their (excess) electricity for profit."[186] But this is Bangladesh—one of the poorest countries in the world.

The technology behind microgrids is about much more than smartphones and the information they offer. However, without that information, on a smartphone or other platform, microgrids would be little different from Edison's first small generating plant, with extremely limited functionality. Net metering laws were created because the current electrical grid is built around large, regulated utilities. Without government representing the owners of individual solar panels, there would be little incentive for utilities to buy energy from these small, distributed generators. With microgrids, consumers can buy and sell outside that century-old system. At this point, most microgrids are not independent of the larger grid, but the barriers to selling excess electricity on the market are lower than ever, reducing reliance on the large utilities. There is power in numbers, and technology is allowing more people to join the market.

REDUCING THE COST OF COORDINATION

Buying pizzas that satisfy everyone isn't that complicated with enough time. What makes it difficult is the high cost of rapidly acquiring the information necessary to find the optimal solution. It is a problem of coordination. The time and effort necessary to coordinate are called "transaction costs." When transaction costs are high, we end up with solutions that are suboptimal, like ordering all pepperoni or cheese pizzas. Most everyone can find something they like, even if very few get exactly what they want. The transaction costs manifest themselves in the size of the gap between the pizza people want and the pizza they get. In the case of electricity, a few large utilities are the pepperoni pizza. They provide basically what you want (predictable electricity at costs lower than you can generate it), but they don't provide many options and the ability to deviate from a defined electricity portfolio is limited. If transaction costs are too high, as in the example where people have only one minute to solve the problem, it can mean the problem does not get solved at all. Without the ability to easily share information about how much electricity they use or generate, a homeowner had virtually no chance of selling their surplus energy to a neighbor.

Smartphones and distributed technology can radically reduce transaction costs and open a new world of opportunities to address environmental problems like making solar power available. What seems so modern was anticipated by Nobel Prize–winning economist Ronald Coase sixty years ago.[187] Two of Coase's key insights help explain the power of smartphones to coordinate people for environmental solutions.

First, Coase recognized that acquiring information has a cost. In *The Nature of the Firm*, he explained why organizations bring some functions in-house and why they contract out for others.[188] If the cost of information is high when efforts are outsourced, such as the cost to be certain that a contractor is doing what they promised, those functions will be brought in-house. Reducing the cost of information was the centerpiece of the app that detected outages in Ghana and the Fishackathon challenge to share the price of fish at various ports. Rather than hire people at each port to report the prices—which would be extremely costly—the app made it possible to acquire the information outside the "firm," or in this case, the "boat."

Gathering information is only part of what it takes to address an environmental problem. It also requires people to coordinate to implement a solution. In *The Problem of Social Cost,* Coase explains that problems where one group is creating pollution that harms another are extremely complex and involve many steps.[189] He writes, "In order to carry out a market transaction it is necessary to discover who it is that one wishes to deal with, to inform people that one wishes to deal and on what terms, to conduct negotiations leading up to a bargain, to draw up the contract, to undertake the inspection needed to make sure that the terms of the contract are being observed, and so on. These operations are often extremely costly, sufficiently costly at any rate to prevent many transactions that would be carried out in a world in which the pricing system worked without cost." He provides the example of a factory that locates near homes and produces smoke affecting local residents. It may be worth it to the company to pay everyone who is impacted for the harm they suffer either for their health costs, to prevent them from being impacted in the first place, or to move. Finding everyone who is harmed and individually negotiating an agreement would be virtually impossible. Coase recognized this and wrote, "There is no reason why, on occasion, such governmental administrative regulation should not lead to an improvement in economic efficiency. This would seem particularly likely when, as is normally the case with the smoke nuisance, a large number of people are involved and in which therefore the costs of handling the problem through the market or the firm may be high." The high transaction costs of addressing air and water pollution are exactly why government was the best choice to address those problems in 1970 when the EPA was created. Since it would be impossible to calculate the impact of an industrial smokestack on every resident in the area, the Clean Air Act set standards that required a reduction in the emissions of every plant. Centralizing authority was an efficient and effective approach.

In other situations, putting the power to make environmental decisions in the hands of a few people can have serious consequences. During an economic downturn, the city council of Flint changed the supply of the city's drinking water to a less expensive source. The new water caused corrosion in the city's lead pipes. Despite repeated evidence that something was wrong, in the form of rust-colored water and a foul smell, staff at the Michigan Department of Environmental Quality told the residents, "Anyone who is concerned about lead in the drinking water in Flint can relax."[190]

The EPA, which also had oversight of Flint's water, was slow to act. In a subsequent investigation, the inspector general of the EPA admonished the regional office, noting that "systems designed to protect Flint drinking water from lead contamination were not in place, residents had reported multiple abnormalities in the water, and test results from some homes showed lead levels above the federal action level."[191] One EPA manager attempted to raise the alarm, noting, "We now have data from yet another independent group that appears to show that the children are in fact, being poisoned."[192] The inspector general's report stated that "the conditions in Flint persisted, and the state continued to delay taking action to require corrosion control or provide alternative drinking water supplies."

Even after the leak of an internal memo from EPA manager Miguel Del Toral outlining the high levels of lead in the water, the agency continued to hesitate. It argued that it had to follow a process of review before it could confirm the memo's conclusions. In her book on the Flint water crisis called *The Poisoned City*, journalist Anna Clark observed that the EPA distanced itself in public from Del Toral's report. "The EPA's repeated defense was that

Del Toral's report was the product of his own research; it hadn't been reviewed or approved by the EPA."[193] The EPA's process is in place to help ensure the science is robust and conclusions are well-founded. It reflects a precautionary approach that recognizes the necessity of using reliable data. It also makes quick action difficult in a crisis. There was discussion of using agency resources to purchase water filters for the city's residents. Nervous it would set a precedent for other communities with lead pipes, one EPA official said, "I don't know if Flint is the kind of community we want to go out on a limb for."[194]

All told, the result was one of the worst water-quality failures in recent US history. Yet when former EPA District 5 regional director Susan Hedman testified to Congress, she argued, "I don't think anyone at EPA did anything wrong."[195] Hedman, who subsequently resigned, pointed to many of the larger problems that contributed to the situation. She continued, "I did not make the catastrophic decision to provide drinking water without corrosion control treatment; I did not vote to cut funding for water infrastructure at the EPA. And I did not design the imperfect statutory framework that we rely on to keep our drinking water safe." Those are all fair points, but they display the limited power of the EPA to protect people from unsafe drinking water. Precaution, lack of authority, and a failure of accountability combined to cause delays that put Flint's drinking water at risk. In Flint, precaution made it difficult to make the timely decisions that were necessary to fulfill the agency's mandate. Due to the high transaction costs of information about water quality and of coordination, the citizens of Flint were largely reliant upon state and federal agencies that faced barriers to quickly addressing the crisis.

Additionally, a regulatory approach is not appropriate for other, distributed environmental problems. For those problems, the cost of enforcement is extremely high, and agencies can't hire enough people to effectively uphold the laws. As Coase noted, "The governmental administrative machine is not itself costless. It can, in fact, on occasion be extremely costly." In the case of electricity and renewable energy, it is not merely the cost to taxpayers to pay for the bureaucracy that oversees the utilities. There is also a cost of regulation that increases the price of electricity. Several states have "renewable portfolio standards" requiring utilities to guarantee that a certain percentage of their electricity is from wind, solar, or other renewable sources of energy. Other states require electricity be generated in the state. These raise costs for everyone—rich and poor. Although there are some programs that help low-income families pay their electricity bill, utilities and government agencies cannot determine the best price to charge each individual family, based on their ability and desire to pay for green energy. Smartphone-connected technology, however, can.

THE BARRIERS TO ORGANICALLY EMERGING SOLUTIONS

Radically reducing the transaction costs of collaboration provides the opportunity to coordinate efforts as never before. Taking advantage of that opportunity will require a change in mindset. Calestous Juma, the great scholar of technology and sustainability, wrote in his book *Innovation and Its Enemies* that "technology, the economy, and the associated institutions coevolve as integrated systems. Change in technology often requires complimentary changes in social institutions."[196] If social and government institutions don't change as technology improves, societies won't effectively harness the power the new technologies offer. During the past fifteen years technology has changed rapidly around the world. Some of the most innovative uses of technology are occurring in developing countries, like Ghana, where institutions are weak and allow transition to a new approach with fewer regulatory barriers. As technology costs decline, these alternatives begin to emerge organically.

In countries like the United States, where institutions and a mindset geared to regulatory oversight are strongly entrenched, these organically emerging approaches are too often weeded out before they can flower. These technologies and customer models are still new, and finding the right approach within the existing regulatory system is difficult. An early attempt at creating micro-grids is an example of the challenges.

A company called Yeloha that connected people to make voluntary purchases of solar power was named one of Fast Company's "World Changing Ideas of 2015."[197] By May 2016, they were out of business. They were simply too far ahead of their time and the barriers were too high. The technology—allowing people to invest in solar panels on someone else's house—was not difficult. The regulatory environment was. Founder Amit Rosner told me, "We tried to change fundamentally how things are done. And when you try to make change in a highly regulated market, it is not impossible, but to succeed it requires a lot of time and in some cases a lot of funding. You need to withstand a lot of resistance and change how people think."

Founded in 2015, the company connected people who wanted to buy solar power with people who have available roof space but may not have the funds to install solar panels. For those living in cloudy Washington State (like me), putting solar panels on the roof doesn't make much sense. Ultimately, it doesn't matter if the panels are on my roof or someone else's any more than it matters whether the electricity I use comes from a natural gas generator in my backyard or one in a neighboring state. By allowing me to pay for solar panels somewhere else, where they can be more efficient, it reduces the price of the electricity, while still adding to the total renewable energy on the grid.

Smartphones were a key piece of making this system work. The Yeloha app allowed buyers to see how much energy their solar panels were generating at any moment and encouraged them to share accomplishments with friends. As far as the technology goes, none of this is particularly complicated. The breakthrough was the scale. Industrial energy users contract with utility-scale electricity generators because they are large enough to keep costs low. Homeowners could not do that because the transaction costs of infrastructure and time made it impossible. Yeloha attempted to change that. Ultimately, because they were the trailblazers, they could not make it work. As Rosner explained to me, the technology outstripped the ability to build new resources. Regulations made it difficult to get the financing to expand their network.

Scott Kessler of the Brooklyn Microgrid expressed a similar frustration. "Regulation is the biggest barrier we face," he said. "In order to sell electricity, you have to be a licensed retailer and there is a lot of paperwork. How do we evolve regulation while ensuring the grid is stable?" Companies like Yeloha, the Brooklyn Microgrid, and others are searching for the right approach. Fortunately, the technology allows innovators to be creative. But regulators have an important role to play too. They need to allow—and encourage—new approaches that put more power in the hands of consumers to choose the type of energy they want. Government regulators are used to applying a standard rule across the board—everyone gets pepperoni pizza and stable electricity prices, with a fixed amount of renewable energy. That predictable approach misses opportunities to help the environment.

After studying the ability of governments to address environmental resource problems, Nobel Prize–winner Elinor Ostrom found that top-down solutions too often fail because the one-size-fits-all approach is unsatisfactory to those who have to actually live with the rules. She noted that "the tendency to try to impose uniform rules throughout a jurisdiction, rather than specialized rules that apply to localities within a jurisdiction, makes it extremely difficult for such officials to set up and enforce rules that will seem effective and fair to local appropriators."[198] This is the point Coase emphasized when he talked about the cost of government and enforcing the laws it adopts. As Ostrom looked around the world she found the best solutions bring people together to create approaches that work for those being affected. These custom approaches often reduce the cost of enforcement because the people at the table can effectively hold each other accountable. Typically, these arrangements involved small, local groups of people. The transaction costs of developing custom solutions for a large number of people were simply too big, which Ostrom said was a "difficult, time-consuming, conflict-invoking process." It can still be difficult, but the barriers are now lower and the opportunities greater.

Microgrids are not the only approach that offers the opportunity to buy green energy on the market. A company called Drift is providing customers with options about the type of energy they buy. Based in Seattle, Drift operates only in New York because of Washington State's strict regulation. Customers buy renewable energy from Drift and each month they get a breakdown of where the energy came from—hydro, solar, nuclear, etc. Drift shops around to find the best mix of renewables so the price remains low.

Drift also puts smartphones to work to coordinate among its members to conserve. When demand was high, Drift lets users know when they should conserve, and then sells the energy. A penny saved is a penny earned, and a kilowatt hour that is unused is less expensive than one generated by a renewable source, while yielding a similar CO_2 reduction. As one article explained, "For now, the process is manual: Building managers are alerted by text message to lower their energy use. As sensors and automated appliances enter the market, Drift plans to allow residential customers to respond to demand events and trade automatically on the energy market."[199] Smartphones make it possible to coordinate, virtually immediately, with large numbers of people to respond to electricity demand.

In each of these situations there is tension among these companies, taxpayer subsidies, and government regulations. Solar power receives enormous subsidies from the federal government. According to the US Energy Information Administration, in 2016 the government handed out $2.2 billion in subsidies for solar energy at a time when it accounted for only 1 percent of the nation's total electricity production.[200] Coal, by way of comparison, received about half that amount while producing 30 percent of US electricity. Wind power received $1.2 billion while producing about 5 percent of total electricity in the US. Whether you support these subsidies or not, it is clear they are reducing the cost of solar compared to fossil fuels and even in comparison to other renewables like wind energy.

Without those subsidies, many of these efforts would be even more expensive, due in part to regulations that make it difficult for new companies to enter the utility market. Rosner complained that when they tried to expand in Massachusetts, regulators set limits he felt were arbitrary, and it became impossible to move forward. Expressing his frustration, Rosner lamented, "Everything depends on an arbitrary decision that you can't predict." Like Drift, which is unable to operate in the state where it is headquartered, state regulations vary and have a difficult time adapting to the rapidly changing technology that is now available to consumers. Regulations are built around the assumption that large utilities will produce energy and distribute it to many small electricity users. When individual homeowners want to act as small "utilities," the system has difficulty adjusting to the new reality.

CONNECTING PEOPLE WHERE THERE ARE FEWER REGULATIONS

Where new markets or opportunities are emerging, smartphones can connect people and create the rules before regulators put barriers in the way. Although electric vehicles like the Nissan Leaf represent a small percentage of cars in the United States, their popularity continues to grow as prices decline.[201] Most EV owners use the car primarily to commute due to the limited range, which allows them to charge their cars overnight while at home. As EVs become more popular and range extends, there may be an increased demand for charging stations in cities and outlying areas so drivers can find them easily. Tesla is building out a network of charging stations to serve its customers.[202] Some cities, hoping to reduce transportation-related CO_2 emissions, are also paying to expand the network of chargers.[203]

Even when drivers can find an EV charging station, they aren't always available. It takes time to fully charge a car, so people must leave their cars at the spot for some time to ensure they have the energy to drive home or run errands. Sometimes the parking spot at a charging station is occupied by a gas-powered car. EV owners call this getting "ICE'd," for internal combustion engine. There is a developing etiquette for EV users at charging stations, but it isn't perfect. Sometimes, disagreements are settled with fists. EV users have a phrase for that too: "charge rage."

Fundamentally, this is a problem of coordination among EV owners. Someone needs to use a charger, but there is another car there. One article described the scenario. "Jennifer arrives in her EV. She parks next to John but doesn't know him or that this is his car. She either has to keep checking on the charging car or leave a note asking the driver to call her when he's done."[204] Even if John doesn't want to hog the charging station, he doesn't know if someone else is waiting and may not know if his car is fully charged. Courtesy alone can't solve the problem. Even though the problem is straight-forward—allow both drivers to communicate so they can share the charging station—it was difficult to connect the two of them. The transaction cost of connecting them prevented a solution.

This is why ChargePoint, one of the largest manufacturers of EV charging stations, created a feature on the app called Waitlist. EV drivers can check the location and availability of charging stations near them. If the station is occupied, users put themselves on a list to be notified when the station is open. Additionally, it alerts the current user that their car is charged, sending a friendly reminder that they should move out of the spot and allow some-one else to use it. Jason Smith of ChargePoint told me, "Everyone gets put in a virtual queue and you get a queue depth that tells you 'one hour.' During that time, if someone else pulled up, the point says it is reserved for driver X. Nobody can cut in line. It is very fair and equitable." The parking spots in front of charging stations may be filled, but the charger itself may not be used for large parts of the day. By allowing users to coordinate, the utilization of a charger increases, making each station more effective. "You are now cycling three or four drivers in a given day," explains Smith, "whereas before it might have been two to three. You are dispensing more energy and getting more utilization." The smartphone was the breakthrough. "Everything is trending toward the phone now," says Smith. "You don't need to have an RFID car. Just tap your phone." The ChargePoint app allows the owner of a charging station to put a price on the time as well. This adds an additional incentive to drivers to stay in the spot only as long as they need to.

It is a nice example of how reducing the transaction cost of coordination can increase the effectiveness of technology and improve environmental outcomes. ChargePoint's app reduces the cost of information and coordination to almost zero. With the click of a button, the app tells the user when the charging station is free. It also tells them when their car is charged. Even though the person currently charging their car and those waiting in line never directly communicate, the ChargePoint Waitlist coordinates their activities to help ensure both of them get what they need while maximizing the use of the charging station. Rather than having to build a huge number of charging stations, the app reduces wasted time at each station, allowing private companies like Tesla, or taxpayers through government subsidies, to spend less and get more. This is an often-overlooked aspect of addressing environmental issues. Every dollar spent needlessly is one less dollar available to fund other environmental efforts. By maximizing the charging capacity of each station, taxpayers can build fewer stations and use the saved money to reduce taxes or address some other environmental issue. That opportunity is created because the transaction cost of coordinating two people is now almost zero, thanks to a smartphone app.

While regulations in developed countries can make it difficult to deploy environmental technologies, the low cost of innovation is making it possible to overcome regulation and corruption that are harming some of the world's most fragile and threatened species.

HOW *BREAKING BAD* IS PROTECTING SEA TURTLES

The leatherback sea turtle is in trouble. In some places, big trouble. To protect them, smartphones may turn out to be more powerful than soldiers with AK-47s.

Leatherback turtles once boasted the largest population of any sea turtle species worldwide. Now they are in rapid decline. With populations in the Pacific, Atlantic, and Indian Oceans, the leatherbacks are not yet in danger of being wiped from the planet, but the trend is exceptionally bad. The International Union for the Conservation of Nature, which maintains the "Red List" of endangered species, lists them as "Vulnerable," with a declining population.[205] That listing, distressing as it is, conceals some dire trends in several parts of the world. In Mexico, for example, the population is now less than 1 percent of its size just a few decades ago.[206]

There is no single cause of the population crash. In many areas, habitat has been destroyed by development. Turtles can become entangled in fishing nets and drown. With specially evolved mouths and throats that allow them to eat jellyfish, they can die after mistaking plastic waste in the ocean for their prey. The primary cause of the leatherback's population collapse in the Pacific, however, is believed to be the poaching of their eggs, which are considered in some places to be an aphrodisiac and can fetch very high prices. One nest can earn a poacher the equivalent of an entire month's salary.

In Nicaragua, the opportunity to poach occurs during a three-day period called the *arribadas* (arrivals) when turtles come to lay their eggs in the sand. As many as 20,000 sea turtles of many different species, including green and olive ridley turtles, show up to nest on Nicaraguan beaches.

When sea turtles lay their eggs, they first dig a hole, deposit dozens of wet, sandy eggs, and begin covering them up. Acting quickly, poachers grab the eggs and put them in a bag. The dark, the sand, the saltwater, and the need for poachers to move quickly are just the opportunity an antipoaching nonprofit called Paso Pacifico needs to protect the turtles.

Paso Pacifico came up with a two-step process to help turn the tables against the poachers. Deep in the clutch of freshly laid eggs, a small, rubbery egg that looks and feels just like a turtle egg—down to the dimple in one end—is placed by a Paso Pacifico volunteer. But it's actually a device called the InvestEGGator that can be tracked by cell phones. When collecting the eggs, Paso Pacifico hopes poachers unwittingly add the InvestEGGator along with the eggs into their bags and drive off. The tracker goes with them, helping conservationists see where the eggs are going to get a sense of the poaching network.

Smartphones make this system possible. "We have this crazy power in our pockets," said Felipe D'Amoed, former director of conservation technology at Paso Pacifico, "but we really haven't used that for conservation."[207]

"Rather than just coming down on the poachers, this is a criminal issue," notes Paso Pacifico executive director Sarah Otterstrom. "The eggs are going to El Salvador and Costa Rica, and we'd like to see governments address the trade," rather than going after the poachers who are often just poor residents.

Experience on the beaches of Nicaragua led Paso Pacifico to target the supply chain. Hollywood inspired the solution. The idea of embedding a tracker in with the bad guys came from the television series *Breaking Bad*. In one episode, a tracker was used to track drugs. But what had once only been available via sophisticated law enforcement devices can now be used by anybody with a cell phone. Andrew Ortner, an associate producer on the show, was delighted to learn of this initiative, saying, "I'm weirdly more proud of *Breaking Bad* potentially helping fight the illegal sea-turtle-egg underground empire versus all the major awards and accolades it's received."[208]

Based in Los Angeles, Paso Pacifico turned to Hollywood special-effects experts to create a decoy egg that felt like the real thing, making it difficult for poachers who are in a hurry to discern the difference. As one report noted, "The artificial eggs are made using a 3-D printer with a material that mimics the look and feel of real sea turtle eggs, which are soft to the touch, not brittle like birds' eggs."[209] The tracker can be picked up where there is cell coverage. The battery lasts for a week and, when detected, location information can be transmitted thousands of miles away. During one of the tests, Paso Pacifico's Kim Williams-Guillen watched on her cell phone, in her home in Michigan, as someone carried the decoy egg around the jungles of Nicaragua.[210]

With just a 3-D printer and smartphones, Paso Pacifico can now fight international poaching rings. This would not have been possible for such a small organization just a few years ago. They still can't be on every beach, but technology is multiplying their efforts. With a limited number of staff, Paso Pacifico needs the technology to deal with the many pressures facing the turtles.

For the impoverished members of the communities where the turtles lay their eggs, the incentives to poach are enormous. Otterstrom appreciates what entices the poachers. "I saw the challenges of poverty that they face," she says. Attempting to balance the pressure of poverty with protection of the eggs, the government "would give 10 percent of the eggs out to the community, but there was a lot of illegal poaching," Otterstrom lamented.

In this existential battle against poaching, the turtles have many enemies—including some who are supposed to be on their side. The Nicaraguan government stations heavily armed soldiers on the beaches to protect the turtles as they lay their eggs. The rewards of poaching, however, swamp the efforts of both government soldiers and park rangers posted on the beach to protect the eggs. Williams-Guillen points out that "even with armed guards, the numbers of poachers overwhelm military personnel by ten or twenty to one."[211]

Since government-sanctioned rangers have the authority to harvest eggs, they sometimes surreptitiously join the poaching. Even if the rangers did want to enforce the law, it is far more lucrative to work with the poachers. Otterstrom adds, "The people who are winning are the government authorities, like the soldiers and rangers, and the middlemen," who ultimately traffic the eggs across Central America.

The corruption doesn't just entice soldiers to join in the plunder, it encourages them to undermine groups fighting to protect endangered turtles like the leatherback and olive ridley. Government officials sometimes receive a cut of the proceeds of poaching and protect the existing system. This makes it difficult for Paso Pacifico to do their work; consequently they avoid government-run beaches. "They don't want us there," says Otterstrom. "It is a threat to their authority. If we did it better, it would make them look bad." The chance of this changing anytime soon is almost zero. In the spring of 2018, demonstrations against the government of Nicaragua turned violent and control over beaches where turtles lay their eggs became even more tenuous. I was scheduled to visit the beaches where Paso Pacifico worked in the summer of 2018 but had to cancel due to the violence. Regardless of whether the government wanted to do something, it doesn't have the ability.

COVID-19 made this situation worse. The economic downturn associated with the pandemic increased the number of poachers. Otterstrom explained that where there were three or four poachers previously, during the crisis there were twenty or thirty. "There are dozens of them," she lamented, "and it will be really hard to get the beaches back from the poachers."

There are other nongovernmental organizations—NGOs—working to protect the turtles, but in the absence of government protection, many of those groups are nervous about taking the risks necessary to fight poaching. As Otterstrom told me, "International NGOs tend to work at a high level, through capital cities, big budgets, and are worried about liability. They aren't in the field, just in the capital." Rather than replacing or pushing to repair dysfunctional government systems, these other bigger and more international NGOs are reliant on the very government officials who profit from poaching.

Otterstrom did not go to Central America intending to help save sea turtles. She grew up in Spokane, Washington, thousands of miles from the Nicaraguan beaches where her work is now focused. With a PhD in ecology, she went to Central America to study the fire ecology of forests and promote conservation. Instead, she saw "the reality of a community that is suffering poverty and a government that is using the situation for their own authority." Watching as thousands of sea turtles came on land to lay their eggs had a profound effect on her. "It is hard not to get passionate," she told me.

Before they created the InvestEGGator, Otterstrom knew that whatever the answer to the problem of poaching might be, it could only be found on the ground. Paso Pacifico begins by acknowledging that local residents have an incentive—even the traditional right—to make money by harvesting turtle eggs. Another Spokane native and former Paso Pacifico employee, Wendy Purnell, who worked in Nicaragua to organize antipoaching efforts, notes that rather than threatening poachers with arrest, Paso Pacifico recognizes "the traditional use of turtle eggs and negotiates with poachers to leave nests intact." In some cases, persuasion works. In others, Paso Pacifico offers cash payments to the poachers.

Ignoring those incentives to poach would be foolish and unproductive. Walking up to impoverished Nicaraguans and lecturing them about endangered turtles likely would not be very successful or welcome. Instead, Paso Pacifico hired former poachers to bridge the cultural gap and make it clear they understood and respected the poachers' motives. This approach worked.

By working directly on the beaches, Paso Pacifico's employees have come to know the poachers personally. Some of them take the cash payments. Others now understand and support the goal of protecting the turtles. "Because there are a few species, green, hawksbill, leatherback, that are endangered, poachers will save the eggs to put in Paso Pacifico nurseries," says Otterstrom. "Some poachers will take half the nest and give half to us for conservation."

Despite overwhelming odds, Paso Pacifico is having success, helping to gradually restore in several areas populations of sea turtles whose prospects were very dim. There are two key factors in their progress.

First, by putting people on the beaches and talking directly and regularly with the poachers, Paso Pacifico is able to personally gather the local knowledge necessary to address the causes of poaching. Government officials sitting in an office in Managua, no matter how well intentioned, simply don't have the same direct and particular knowledge.

Second, Paso Pacifico understands the incentives poor Nicaraguans have to take the eggs. For poachers—not to mention underpaid soldiers—the value of one nest of eggs is too much to ignore. Given a choice between several months' salary and the theoretical impact stealing those eggs causes to worldwide sea turtle populations, the calculus isn't difficult.

However, turtle populations overall continue to decline. Impoverished poachers and their partners in government reap relatively huge financial rewards. They are supported by international trafficking rings that sell the eggs to buyers in several countries. The international NGOs who are supposed to protect the turtles are afraid to take the risks necessary to give them a fighting chance.

Counteracting these forces is a major challenge, and the cost to protect just one beach is significant. Paso Pacifico doesn't have the funding or staff to protect the many other beaches where poaching is still largely unchecked. "We've seen improvement where we work. There is a huge problem everywhere else," says Otterstrom.

The need to magnify Paso Pacifico's impact beyond the few beaches they can patrol was the reason they turned to smartphones and small technology. They hired Felipe D'Amoed to find new ways to use technology to augment their work on the ground. "The interesting thing about the artificial eggs is that it gives us a different strategy to fight poaching," said D'Amoed. Otterstrom adds, "Technology is a force multiplier in more than one way. We can influence other beaches where we don't have rangers."

Smartphones and the internet provided the mechanism to fund the egg's creation as well. Although initial funding came from the US Agency for International Development through their Wildlife Crime Tech Challenge,[212] Paso Pacifico also used crowdfunding to finance additional eggs.

There is still skepticism. After Otterstrom explained the concept behind the InvestEGGator and how it would complement existing efforts, one of Paso Pacifico's field directors said, "This is not going to work," believing that only people patrolling the beaches can make the difference. Otterstrom said getting environmentalists to appreciate the power of technology can be difficult. The goal of the technology is not to replace people on the ground, which would be the environmental equivalent of replacing live operators with the frustration of automated phone systems. The goal of environmental technology like the InvestEGGator is to give people on the ground more tools. "Technology is an intersection between the public and the environment in terms of conversation. A new point of engagement," she said.

Engagement can be a cheap word. I have seen dozens of projects that claimed they would "raise awareness." Almost all of these efforts were lucrative for the individuals promoting awareness yet did little to transform environmental reality. But smartphones and connected environmental technologies can help change that. Connecting people to real activity on the ground is far more meaningful than adding a name to an online petition or sending a tweet.

The InvestEGGator is starting to be meaningful. Partnering with Helen Pheasey, a PhD student at the University of Kent in England, Paso Pacifico tested 100 of the decoy eggs in Costa Rica where the political situation was more suitable. With a group of volunteers, Pheasey would walk along the beaches at night looking for turtle tracks and then follow them to a nesting site. After finding a turtle in the process of laying eggs, the volunteers would put on a glove and place an egg in the middle of the clutch. "It was as simple as that," said Pheasey. "The long bit was finding the turtle." It was more complicated on the Atlantic coast where they were protecting green turtles. "They poach the green turtle for the meat," she explained, "so you have to stay with the turtle for five to seven hours" to make sure they return safely to the ocean.

Eggs were tested on both the Atlantic and Pacific coasts and on a number of different species. The InvestEGGator is about the size of a green turtle egg and Pheasey wondered if poachers would notice when it was inserted into a clutch of olive ridley eggs, which are much smaller.

There were many failures. Most of the eggs were never poached (Pheasey and her team marked the locations so they could go back and retrieve them later). Sometimes the poachers figured out what was happening. The first time one of the eggs was poached, she opened the tracking app to see where it was. "It was the first one I had, and I was bouncing off the walls," said Pheasey. "We looked at the track and it was inside a riverbed," where a poacher had thrown it. In another case a poacher took the egg apart and sent photos to Costa Rican antipoaching NGOs asking what it was. "We said it was temperature testing," said Pheasey.

Ultimately, her persistence paid off. Despite planting dozens of eggs, none had produced a complete track from the nest to a market and Pheasey was getting frustrated. "It was a nightmare." On the last night of the effort, she was tired and almost didn't go out. As luck would have it that was the night that yielded the best results. "One nest got poached and it was the one with my egg. It was a joyous night," she said, noting she felt strange celebrating poaching. They tracked the egg as it moved across the country, finally stopping 137 kilometers from where it was deployed. "I started looking at the map and it is a supermarket, and they were at the loading bay. There is no reason for the egg to be at that place," said Pheasey. "I drove there myself and it was a dodgy back alley." Later the egg moved to a residential area where poachers will sell the eggs door to door as a snack. "They are not fulfilling people's protein need in an impoverished situation. They are a bar snack. They are a treat item."

The experience proved the viability of the technology. The egg was durable enough and had enough battery power to be tracked across the country, exposing the entire network, from poacher to buyer. Pheasey believes there is potential not only for sea turtles but for other wildlife crimes. "Other countries have problems with poaching of crocodile eggs. You'd have to change your strategy about deploying them," she said with a laugh.

Otterstrom says that the eggs have been sold to a South American country as part of their antipoaching campaign. "There is a group in Kenya who wanted to buy some and someone in Malaysia reached out to us about buying some eggs. There is definitely global interest," says Otterstrom.

Paso Pacifico is a great example of the opportunity for small technology to create environmental improvement where government-led efforts have failed or even hindered progress. The story of the InvestEGGator highlights the power of local knowledge, coordinating the actions of many people, and using personal incentives for the benefit of the environment.

COORDINATING FOR CONSERVATION

In some cases, governmental regulation is the best option because the other choices are ineffective. From the standpoint of preventing extinction, the Endangered Species Act (ESA) has been a success. From the standpoint of recovering species, it has not. The transaction costs of coordinating private landowners, who have the land necessary to provide the habitat needed by endangered species, are very high. Regulations can help preserve remaining habitat but are less suited to facilitating the coordination that is necessary to increase the pool of available habitat necessary for recovery. This trade-off has been widely acknowledged, but until now there were few reasonable alternatives.

By preventing landowners from using their land, the ESA makes listed species a liability rather than an asset. That gives landowners an incentive to destroy habitat before endangered species move in, bringing land restrictions with them.[213] Jonathan Adler of Resources for the Future wrote on the thirty-fifth anniversary of ESA that "landowners have been known to destroy or degrade potential habitat on their land preemptively in order to prevent the imposition of the act's requirements. It is not illegal to modify land that might become endangered species habitat some day in the future, nor are landowners required to take affirmative steps to maintain endangered species habitat beyond refraining from actions that 'harm' endangered species."[214] Ironically, ESA creates the perverse incentive to destroy habitat rather than protect it. This is particularly problematic because, as the US Fish and Wildlife Service notes, "Two-thirds of federally listed species have at least some habitat on private land, and some species have most of their remaining habitat on private land."[215]

There are taxpayer-subsidized programs designed to compensate landowners for the value they've lost. The problem is that many of these efforts are underfunded, and landowners are often skeptical of government promises to pay. In Washington State, families who own small amounts of forestland are eligible to recover part of the lost value of timber along streams that are now required to have buffers. Funding for that program, however, is scarce and there is a long list of families waiting to receive compensation. Some, knowing money isn't going to arrive any time soon, simply haven't applied.

One basic problem is that those who have habitat worth saving can't find those who might be willing to pay to protect it, whether it is restricted by the ESA or not. Just as the Brooklyn Microgrid helps neighbors find others willing to sell them electricity, or Yeloha helped people without a good location for solar panels find others who had rooftops in sunny parts of the country, the same could be done with landowners and those who want to preserve wildlife. The pop-up wetlands created by the Nature Conservancy using eBird is a prime example of what can be done to identify and pay for habitat protection. It is not the only example.

There is already a market for protection of wetlands, where developers who need to offset their impact on habitat can invest in protection or creation of habitat nearby. Creating habitat this way, known as "mitigation banking," can be a win-win for the economy and wildlife. Describing the program, one government agency noted that "wetland mitigation banks are more likely to succeed than permittee-responsible mitigation projects because the banks are built before damage occurs to another wetland site. Banking has the potential to increase ecological benefits, save money for project applicants, and make the application and permitting processes more efficient. Banking also creates an economic incentive for restoring, creating, enhancing and/or preserving wetlands."[216] This system is set up for large businesses to connect with companies that create habitat to satisfy regulation. This works well because even though there are transaction costs when coordinating a buyer and seller, the size of the purchases makes those costs relatively small. For those with small amounts of money the transaction costs are prohibitive. Smartphones make it possible to democratize habitat restoration, reducing the cost of connecting donors who want to protect habitat and landowners who have habitat worth protecting.

Directly contributing to wildlife habitat in this way can also address a concern that nags at many who donate to organizations promising to help the environment: how do you know they helped?

Environmental groups raise huge sums of money to change public policy, but it is difficult to measure whether they are successful. If a policy they advocated didn't pass, does it mean they did a bad job, or does it mean they made progress toward improving public policy in the near future? I've worked with a number of nonprofits, including those that focus on changing policy, and sometimes even we have a hard time deciding if we made any progress.

Rather than contribute to a political group in the hope it will change policy, environmentally conscious individuals can help increase the amount of salmon habitat, for example, by paying to create wetlands. Zach Woodward, who has been building wetlands for many years, says people could buy habitat for a certain number of Chinook salmon. "I could say that a certain number of ESA Chinook salmon credits are being held" for people who want to contribute to habitat that saves them.

It can also overcome a fundamental challenge with using private land to protect wildlife habitat. The benefits of wildlife habitat and endangered species are enjoyed by everyone. They are a societal good. Even if we never see a dusky gopher frog or a gray wolf, we feel better knowing they are there. Frequently the cost to protect those species is borne by landowners who happen to be in the wrong place. We could argue they should be proud to be part of species recovery and argue they don't have the right to harm endangered species, but the corollary is that if society benefits from healthy species, society should pay their share. Sometimes this works. Sometimes it does not. The distance between what is ideal— governments having the resources and willingness to do their part for species recovery—and reality can be filled by private donors, large and small, who care enough to offer their own resources. Private contributions to conservation aren't new, but they have been limited due to the difficulty of connecting donors to projects, especially those who have small amounts of money to contribute. When the cost of collaboration goes down to almost zero, the ability to democratize species recovery could grow significantly.

This is the magnifying power of collaboration. Neighbors can trade electricity. Drivers can maximize the use of charging stations. People who have never met each other can provide habitat for endangered species. In some ways what smartphones and information technology have done is improve what was already occurring. This could be dismissed as incremental improvement. It is more than that. Driving the transaction costs and the cost of collaboration to almost zero helps overcome the problem that Ronald Coase won the Nobel Prize for identifying. The reason governments act as surrogates for citizens is because the cost of collaboration among hundreds of millions of people, or even tens of thousands in a city, is too large. Government taxes and regulations may be expensive, but the alternative was even more so. In the case of endangered species, the cost is too high for government, with only a very few species ever achieving recovery. By making personal contributions to recovery feasible, we add a new tool to help ensure a future for creatures like the gray wolf.

As more people generate and sell electricity and contribute to species recovery, that isn't just an incremental improvement. It can be a game changer.

Enhancing Our Connection to Nature

Steelhead trout, a migratory fish that lives primarily in the western United States, face a dangerous trip from the streams where they are born to the ocean where they will live most of their life. In the Puget Sound on the coast of Washington State, they must survive a nearly two-hundred-mile journey, avoiding predators like seals, sea lions, and birds. They navigate past man-made obstacles like floating bridges that narrow their passage and can confuse the young fish. There might not be enough food on the route to keep them strong as they grow. As a result, only about 20 percent of juvenile steelhead successfully make the trip.

Helping more steelhead reach the ocean requires that we understand these challenges. A group called Long Live the Kings—a play on the name of the endangered king salmon they are working to recover—used small technology to track the fish and assess the obstacles they face. The organization inserted a tag in the belly of steelhead released in rivers that flow into the Puget Sound. The tag communicates with 125 sensors placed throughout the sound, tracking them along their journey. The data about the routes the fish take is useful to diagnose what is causing increased mortality.

167

"Steelhead maintain a relatively consistent speed and trajectory as they race to the Pacific Ocean," the organization explained on their web page.[217] "When tracking data show that they've deviated from that path or speed, it means they've likely encountered some sort of obstacle during their migration. The obstacle could be a physical barrier, an encounter with a predator, a decline in health related to disease, and/or a variety of other things."

The research was so illuminating, Long Live the Kings decided to turn the fish tracking into a game called Survive the Sound, giving the public and schoolchildren the opportunity to choose a fish, each with a unique name, and follow its fate while learning about the science of salmon recovery.[218]

As a member of the state Puget Sound Salmon Recovery Council, I was one of the first to hear about this innovative approach. My important contribution to the project was to name one of the fish "Fishy McFishface." Mock the name if you like, but in my defense, when introduced to the second-graders who were the target of the lesson, the *Seattle Times* reported, "A cartoon juvenile steelhead named 'Fishy McFishface' is stealing the show."[219] Vindication.

Exposing students, or anyone concerned about the survival of salmon, to the challenges they face is important not merely because it raises "awareness" (a term vaguely used to justify all sorts of silly environmental expenditures) but because it creates a connection between people and the real-world problems scientists, activists, and policy makers face. It is easy to argue we aren't doing "enough" to help salmon or the environment. It is more difficult, and more important, to be clear on what should be done by creating a depth of understanding that isn't always present in political discussions.

In the case of Puget Sound steelhead, the data from the tracking revealed a particularly vexing reality: a major obstacle to increasing salmon populations is the very large population of seals and sea lions. This problem is widely acknowledged by salmon scientists but is less appreciated by the general public. There are many reasons salmon and steelhead are not recovering, and the blame can't entirely be pinned on hungry marine mammals. However, they do play a role and it is part of the complexity of the issue that piques the interest of people, drawing them in with counterintuitive information that helps turn an interest in the environment from a general concern to something more robust and serious.

BUILDING THE STEWARDSHIP ETHIC

Helping people appreciate the complexity of environmental challenges is critical to building an ethic of serious environmental stewardship that is rewarding and contributes to effective environmental protection. Personal technologies offer a remarkable new range of options to comprehend ecosystems and wildlife. But the value of democratizing participation in environmental science has long been appreciated.

Aldo Leopold, who is often called the father of conservation, highlighted the benefits of engaging nonscientists in the understanding of natural systems. In his 1943 essay "Wildlife in American Culture," Leopold noted that while the preservation of wildlife was an important part of the national ethos, the actual work of understanding the science was limited to a very few people.[220] He called for expanding wildlife research to a much larger audience, engaging people who may feel their lack of expertise or credentials prevents them from contributing. He called these efforts "wildlife sports." His description of these "sports" sounds even more appropriate today. This "totally new form of sport, which does not destroy wildlife, which uses gadgets without being used by them, which outflanks the problem of posted land," he explained, "needs teachers, but not wardens." The gadgets have become more sophisticated—as we will see— but his argument is even more compelling eight decades after he first made it. The more meaningful the role people have in contributing to wildlife science and preservation, the stronger their connection to conservation. "In my opinion," he wrote, "the promotion of wildlife research sports is the most important job confronting the profession of wildlife management."

He wasn't the only one to recognize the value of building a connection and sense of ownership to environmental stewardship. In his book, *How to Think Seriously about the Planet*, philosopher Roger Scruton argued in 2012 that a love for your home and a sense of place—what he called "oikophilia"—was the prime motivator of a conservation ethic and an important impetus for effective environmental stewardship. When faced with environmental degradation, "the best thing is that ordinary people, motivated by old-fashioned oikophilia," Scruton argued, "should volunteer to localize the problem, and then try to solve it."[221] A tangible connection to environmental conservation is not only effective but also creates a bond and a sense of responsibility that is conducive to good stewardship. Scruton offers the example of Ravenna Park in Seattle, which was a private preserve of giant old-growth trees. Concerned about the protection of those trees, the city of Seattle purchased the land. Not too soon afterward, some of the largest trees began disappearing as the city decided that older trees were a safety hazard. When the stewardship ethic was diluted with other priorities—valid or not—the protection disappeared and trees that had survived for hundreds of years were lost.

When concerned citizens feel a disconnect between the desire to protect the environment and the opportunity to act as good stewards, it also makes good public policy more difficult. I have seen the frustration that results from decoupling concern for the planet and participation in the solution. Working at a government environmental agency, I helped develop environmental impact statements (EIS), some quite large. I came to believe that those long and comprehensive documents were more about show than actually involving the public. Feeling sidelined by the detail and jargon, many people decided the only alternative was to show up at public meetings and yell. I have witnessed many public comment sessions where people dress as killer whales, sing songs, and engage in other such behavior that doesn't add much to the rational scientific discussion that is supposed to be occurring. Even as I sometimes rolled my eyes at the antics, I also came to realize that many of these things occurred because the barrier to more meaningful participation was very high. Few people have the time to read, and then later critique, the details of an EIS hundreds of pages long. Anyway, who would trust a random dude over the word of a credible scientist?

Enabling a more robust stewardship ethic, connected to real-world problems and empowered by technology that makes action possible and meaningful, offers the opportunity to move beyond environmental performance art to true engagement. The "wildlife sports" that Aldo Leopold identified eight decades ago are more feasible and essential than ever.

HEARING THE RAINFOREST (AND ILLEGAL LOGGING)

Few people in Europe and America will have the opportunity to enjoy the sounds of tropical rainforests around the world in real time. With a click on an app, now anyone with a smartphone can do just that—and it helps fight illegal logging and deforestation. The technology that makes this possible is called, appropriately enough, Rainforest Connection.

Topher White, the founder of Rainforest Connection, said what struck him about the rainforest he visited in 2011 was the noise. There is a constant hum of sounds, including birds, insects, and in the place he visited, gibbons, a type of ape that lives in Asian rainforests. By downloading the RainforestCx app, you can hear many of these sounds recorded using very simple technology. By connecting an old cell phone to a few small solar panels, White and a few others created a system that would listen to the forest and send the sounds to the internet, where they could be downloaded and heard by anyone. Even in the middle of the jungle, the cell coverage was sufficient. "There was no electricity, no running water, no roads, but there was a pretty good cell phone service," said White.[222]

The app has several recordings, including sounds of insects, birds, monkeys, frogs, and rainstorms. Or people can listen to live sounds from locations in Romania or Malaysia. As I write this, I am listening to a thunderstorm in real time in Similajau National Park in Malaysia. I was a little skeptical, so I checked the weather over Similajau and, sure enough, it showed a thunderstorm in progress. The app even allows you to save a particular moment that is interesting, like a unique bird song.

As entrancing as the sounds of the jungle are, that wasn't what motivated White to create the technology. He was listening for the sounds of chain saws that might reveal illegal logging. The problem is widespread, and according to Interpol, "Tropical deforestation accounts for 10–15% of global [carbon] emissions, and nearly 50–90% of the logging is illegal in major tropical countries."[223] As White told *National Geographic*, "detecting chainsaws and other sounds related to that activity can be tough, because the air is already filled with the cacophony of nature."[224] Using artificial intelligence, RainforestCx analyzes the sounds collected by the cell phone/solar panel devices, which they called "Forest Guardians," and sends an alert to local rangers when it detects chain saws, gunshots, or vehicles. The device is powerful enough to hear sounds a mile away.

The impact of the listening devices was almost immediate. On the second day of installing Forest Guardians in the treetops, White received a notification that the devices detected chain saws not far from where they were. Pausing to listen, the group heard the faint sound of the saws in the distance—noise that had been obscured by other natural sounds. The group set off in the direction of the chain saws and caught the illegal loggers in the act, ultimately convincing them to stop.

The Rainforest Connection team built on that experience, improving the quality of the data to help those on the ground take action. Chrissy Durkin, director of international expansion for Rainforest Connection, says the goal is to tailor the information to the needs of their partners on the ground. "The more data and understanding we can provide to them," she explained, "the better informed they can be to make smart decisions about how to approach" suspected illegal logging operations.[225] How big is the operation? How many officers do they need to bring? "Having that data is important to the safety of our partners," she says. A group of lawyers is using the sound data as forensic evidence to prosecute illegal loggers.

The experience gained developing the Forest Guardians and analyzing the sounds has the potential to help in other areas of conservation. "If we build the system well, so people can really pull out some amazing insights, we can create a market for bioacoustics that does not exist right now," White explained in one interview. "We're on the verge of a pretty exciting moment, almost like the invention of the microscope, when you're able to connect AI with large-scale bioacoustics in order to look into elements of nature that no one was aware existed." By collecting acoustic data from forests in several parts of the world, Rainforest Connection is creating a repository of sounds that can be used to stop illegal logging as well as contribute to conservation efforts in these critical ecosystems.

"Acoustic data is particularly rich in information about biodiversity," says Durkin. "Most animals make noise." Using artificial intelligence, computers can learn to identify animal sounds, providing information about where and when animals are in parts of the forest. Using the streamed data—like the kind I was listening to—Rainforest Connection is creating a biodiversity database. The ability to identify animal sounds provided by the Forest Guardians in real time turns them into, in Durkin's words, "weather stations for biodiversity." By identifying locations where there are important species, conservationists can target their efforts and protection. And, by tracking the sounds over time, they can track the impacts of their efforts. Because the information is accessible, it can be used by small groups with few resources. "It takes the scientist out of the equation," says Durkin, removing a cost that might otherwise be incurred. By democratizing access to high-quality information, Rainforest Connection reduced a barrier that made effective conservation challenging.

That breakthrough is being driven, in part, by advances in the quality and affordability of listening technology. (And, just to make it the perfect example of the power of small environmental technology, it was also financed through a crowdfunding campaign on Kickstarter.)[226]

Initially, the Forest Guardians were made from "upcycled" cell phones donated by people who wanted to contribute in a small way to help protect the rainforests. New versions use custom technology to ensure the durability necessary to survive in the rainforest environment for multiple years. The number of inexpensive, open-source conservation technologies is growing. Alasdair Davies, cofounder of the Arribada Initiative, which creates affordable conservation technologies, told the *Wall Street Journal* that these improvements are critical to expand opportunities for a community that is often short of resources.[227] Researcher and conservationist Lydia Gibson, who uses open-source technology in her conservation work in Jamaica, told the paper that "open-source technology is also allowing more people to engage in environmental research."

Survive the Sound and Rainforest Connection are effective at bridging the gap between people and the environment because they do more than just provide the ability to experience something. Both of them combine connection with meaning. It is the difference between enjoying a beautiful grove of trees and feeling a sense of responsibility for their stewardship. It is why people who heard about Rainforest Connection started sending their old phones to contribute to the effort. It is the conservation ethic that Leopold and Scruton highlighted. Indeed, Scruton argues that providing more opportunities to connect are critical to building that ethic. "If they are losing the habit of doing this," he wrote, "it is in part because governments, responding to pressure groups and activists, have progressively confiscated the duties of the citizen." Sometimes, when government action is the best option for environmental protection, the loss of opportunities to build that ethic is worth the price. In the case of illegal logging, establishing an emotional connection to the sounds of the rainforest and a stewardship ethic to protect what you hear is extremely valuable.

I know this is true because it is how I started my career in environmental policy. When I began work at the Washington State Department of Natural Resources, I knew little about forestry science and the factors that impacted the health of forests, wildlife habitat, stream protection, and the myriad other trade-offs involved in forest management. After spending time in state forests, listening to biologists, silviculturists, and field foresters, I began to discover things I'd never heard in the media or public discussion. I was intrigued. I see that same dynamic at work with Rainforest Connection, Survive the Sound, and numerous other opportunities that allow people to see nature in a way that is new and captivating. That is the spirit behind the creation of one of the most popular environmental apps for citizen scientists: iNaturalist.

EVERYONE CAN BE A WILDLIFE SPORTS STAR

When the iNaturalist app was released in 2008, the goal was straightforward. For creator Ken-ichi Ueda, who conceived the idea as part of his master's degree at the University of California at Berkeley, it was "really about engaging people with nature through technology."[228] Using artificial intelligence, iNaturalist allows anyone to take a photo of a plant, insect, or animal and the app will attempt to identify it. In many instances the artificial intelligence works and provides a ranked list of potential identifications, with photos to help match. Sometimes it will provide only the likely family. In other cases, the identification is exact. The app also shares the photos with the iNaturalist community for verification or additional identification.

The AI that drives iNaturalist is based on recognizing patterns in photos provided by users. By analyzing photos of species taken from several angles, the AI can identify distinguishing characteristics and create a classification. Cornell University research engineer Grant van Horn began working on computer vision in 2010, developing methods of using photos to identify birds. As more photos were included, the system improved. "Large data plus large computation equals results," van Horn explained.[229] "If you show the model a thousand different photos of a moth species, it is very unlikely you are going to take a photo that it hasn't seen." As the bird identification system became more accurate, van Horn and his colleagues met with iNaturalist to apply their model to a wider range of species. The fundamental challenge is the same—identify patterns unique to a species, whether that is a bird, a plant, or an insect. The photos provided by users became more important to ensure there was an adequate amount of information for the AI to calibrate. When there are a thousand photos, recognition is easy. When there are only ten, the rate of successful identification declines significantly. The threshold for reliable AI identification is about twenty photos. In a research paper tackling the challenge of identifying species with few photos, van Horn and his colleagues noted that the growing number of iNaturalist users was rapidly expanding the data set for identification. "Currently, every 1.7 hours another species passes the twenty unique observer threshold, making it available for inclusion in the dataset," they wrote.[230] There is a symbiosis between information provided by citizen scientists and machine learning. The more effectively iNaturalist and other citizen science platforms can engage people, the better the systems become. "Cornell stresses the importance of putting a name to something just so you can tap into all the additional information about the organism," says van Horn. "That is what gets people excited about a species and opens the door to engagement." That model of using information to engage people has been very successful.

One of iNaturalist's founders, Scott Loarie, explained that the goal was to connect people to the environment. Rather than creating a tool that identified what observations were needed, iNaturalist's creators focused on making it good for the user. "How do we offer a tool that people want to use in their own ability to connect with nature?" asked Loarie.[231] "The effort is important for science and conservation, but our strategy has been on building the community and the incentives to participate." Traditional citizen science efforts direct participants, providing information about what information they want collected. iNaturalist allows people to submit information they find interesting. "That is one thing that distinguishes iNat from others. It is self-directed," Loarie told me. The usability increased the number of sightings. iNaturalist's outreach coordinator Tony Iwane started as one of the testers in its early stages and said he became hooked. "Once you start noticing a few things, you want to find more things," he said. "When I first started going out, I just wanted to find spiders, reptiles, or amphibians. But now I photograph plants and you start noticing everything else. I don't want to say it is an addiction, but it opens your eyes."

Others enjoyed a similar experience. Iwane recounted one story of a hiker in the San Gabriel Mountains near Los Angeles who posted a photo of a snail on the app. It turned out to be a snail unseen in seventy years that had been considered extinct in the area. Researchers saw the photo and did some additional research, which was later used to petition for protection of the snail as an endangered species. "The person who took the photographs is now one of our top snail guys in iNaturalist," said Iwane. "He is now studying snails and it changed the course of his life. This is the ideal. [iNaturalist] brings together anyone who is interested in nature." This story is remarkably similar to a description from Leopold about the way citizen science helps deepen an appreciation for the environment, taking people from casual enthusiasts and making them experts. In "Wildlife in American Culture," Leopold offered some examples:

Thus Margaret Morse Nice, an amateur ornithologist, studied song sparrows in her backyard. She has become a world authority on bird behavior, and has out thought and out worked many a professional student of social organization in birds. Charles L. Broley, a banker, banded Eagles for fun. He discovered a hitherto unknown fact: that some Eagles nest in the South in winter, and then go vacationing to the Northwoods. Norman and Stewart Criddle, wheat ranchers on the Manitoba prairies, studied the fauna and flora of their farm, and became recognized authorities on everything from the local botany to wildlife cycles. Elliot S. Barker, a cowman in the New Mexico mountains, has written one of the two best books on that elusive cat: the Mountain Lion. Do not let anyone tell you that these people made work out of play. They simply realized that the most fun lies in seeing and studying the unknown.

iNaturalist has opened those opportunities for many more people by breaking down the barriers between casual observers and the experts. With a photo and a click, people can learn more about the natural wonders that surround them, with the assistance of a community of experts, called the "iNat community," who offer education about what they see.

The technology has so thoroughly blurred the line between amateur and scientist that anyone's casual photo of a snail or insect can be considered "research grade" with a little additional effort. Photos posted on iNaturalist must first be considered a "verifiable observation," which simply means the photo has a location, a date, and isn't of a captive animal.[232] The app does the work of attaching a location and date to the observation and sends it to the iNat community for additional verification. If members agree with the identification, it becomes "research grade." Not every observation can become research grade. Some species require more information, even when a photo is sharp and clear. Photos or sounds that have been classified as research grade are then shared with the Global Biodiversity Information Facility (GBIF), an open-access database of information about biodiversity and species.[233]

The success of iNaturalist has been remarkable. By the end of 2020, more than 1.3 million users had entered 50 million plant and animal observations.[234] There are an estimated 2 million species on the planet, one-sixth of which—more than 333,000 species in total—have been tallied on iNaturalist.[235] The California Academy of Sciences and *National Geographic*, who became the sponsors of the app in 2014 and 2017 respectively, have now set a goal to reach a million identified species by 2030. It has also become an extremely successful source of scientific information. GBIF tracks more than 60,000 data sets of species information, and when a researcher downloads any of those sets to write a paper, GBIF staff can track which data were used. iNaturalist has the largest number of citations. The identification system has become so robust that the iNaturalist team launched Seek, an app that can recognize species in real time, just by pointing your phone's camera.

As impressive as those results are, it is still limited to just one aspect of the environment—identifying plant and animal species. There are many other areas where citizen science plays an important role.

BREAKING DOWN THE BARRIERS

By making data collection and information sharing simple, smartphones have opened science up to entirely new participants—people who thought they couldn't contribute to important research. A report from the National Academies of Sciences, Engineering, and Medicine on how researchers can engage citizen scientists notes that "citizen science allows people with diverse motivations and intentions to participate in science."[236] Allowing nonscientists to collaborate in interesting and important scientific questions is what motivated Darlene Cavalier to create SciStarter, a hub for more than 3,000 citizen science projects and 110,000 citizen scientists.[237]

When working at *Discover* magazine, Cavalier said the programs she organized for school children were designed to encourage kids to become scientists. Cavalier, however, didn't have a science degree and felt the message was leaving out people like her. "It didn't work for me," she said. "What do we do with all the other people, people who don't have formal science degrees?"[238] Citizen science was a way to include others who wanted to contribute in their own way.

In the beginning, SciStarter was just a WordPress blog with a database of projects. After five years, it took off when the National Science Foundation provided a grant to help connect scientists with people wanting to contribute to our understanding of the impacts of climate change. Researchers put their projects on SciStarter, and individuals or groups looking for a way to contribute can find something that interests them. Projects cover a wide range of topics, from the environment to health research. Some require participants to get into the outdoors, tracking birds, identifying monarch butterfly habitat, or finding jellyfish, for example. Others are online and ask participants to interpret video or images. Some are focused on a nearby project; others have global impact. "It creates onramps for people who are curious or concerned about a variety of topics," explained Cavalier. The immediate goal of citizen science is to accelerate research. For participants, however, there is another benefit. "Engagement in citizen science has been shown to increase confidence and help create an identity as someone who can and should participate in science," Cavalier said. "It breaks down the barriers between the research community and those who want to apply science to solve problems."

Projects like SciStarter are successful, and an immense amount of data is provided by citizen scientists. There are, however, doubts about the reliability of citizen science data. Scientists who rely on data collected by members of the public have systems to confirm the quality of the information, filtering out information that appears anomalous. More data isn't necessarily better if it is low quality. Poor data leads to poor research, or as researchers say, "Garbage in, garbage out." That's exactly what the scientists behind the Marine Debris Tracker, an app built in conjunction with *National Geographic* and researchers from the University of Georgia, are trying to avoid. Cavalier says it is one of her favorite citizen science projects.

Although there is a growing concern about the increase in the amount of plastic and trash in the ocean, researchers are still trying to understand how trash moves from cities into the ocean. Solving that problem requires more data than can be gathered by individual researchers. To make up for the shortage of data, Kathryn Youngblood helped develop the Marine Debris Tracker to engage the public in finding and reporting trash in key cities along the Mississippi River. Youngblood, an environmental engineer at the University of Georgia and citizen science director of the debris tracker program, worked with groups to help collect data in those communities. "Plastic pollution seems to be this far-away problem," she told me.[239] "Collecting the data in the area around you is a way to get people involved. . . . It is an issue we all see in our day-to-day lives, and we all have some responsibility or have some awareness about how things are ending up in the environment."

Engaging volunteers is necessary if she and her colleagues are going to find ways to track, and ultimately prevent, plastic pollution from reaching the ocean. A short manual for volunteers in the Mississippi River study explains how important their role is. "While we know the plastic pollution problem doesn't start where the river meets the Gulf of Mexico, we don't know the what, why, or how of litter entering the river from our communities," it explains.[240] "We need community scientists to help us find the missing piece of the puzzle."

The Marine Debris Tracker app was initially developed as part of a *National Geographic* effort in India to find the sources of pollution in that country. The app makes it easy to report trash while including safeguards that ensure the information being entered is reliable. When someone finds a piece of trash, they can choose one of the preset categories, which include aluminum cans, balloons, tires, glass fragments, plastic bottles, and more than fifty other categories. Users enter the number of items and take a photo. The app uses the phone's GPS to identify the location. Youngblood says they do some quality control in the app, including preventing someone from entering a very large number of items. The photo also helps with data verification. Ultimately, she says the quality is very good. "Litter tracking is not rocket science," she said. "Most people who take the time to do citizen science are doing the best they can."

Rather than simply relying on individual members of the public to provide data, her team also worked with groups in St. Paul, St. Louis, and Baton Rouge to provide guidance and training and improve the quality of the data. "We had folks on the ground doing urban transects where we randomly chose places in the city," said Youngblood. "We had people in accumulation areas, along the riverbank. And the third thing we had them do is look at floating debris along the river." By instructing volunteers where to look for trash, researchers increase the likelihood that data are from a representative distribution of locations in each city. After a month, more than 75,000 items had been logged into the app.

Researchers also put small technology to work supplementing the data from citizen scientists. Much like the GPS trackers used in the InvestEGGators to track poached turtle eggs, the Debris Tracker team put bottles with GPS trackers in the Mississippi to see how trash moved down the river. "We deployed one in St. Louis and had it returned in Baton Rouge two weeks later," said Youngblood. In another instance, a bottle became stuck in some branches and then suddenly began moving upstream. The researchers believe it may have become snagged on a barge.[241] The information from the tracked bottles supplements data collected by citizen scientists, adding depth to the information.

Key amongst the findings is that the type of pollution is location dependent. In the Mississippi River, plastic bottles were common. In India, however, plastic bottles were extremely rare. Even though there isn't an effective government-run system of recycling, there is an aggressive, private-sector organization to collect plastic bottles, paying people who return them.

Collecting a large amount of quality data from citizen scientists, supplemented by geographic tracking of plastic bottles, is far easier now than it was a decade ago. Most people have a smartphone and can participate casually even if they are unable to be part of the more organized effort to log the location of trash. The ability to track debris isn't new, but the availability and sophistication of the technology have improved dramatically. It is so ubiquitous there are some unintended consequences. Youngblood told me that after someone found one of the tracked plastic bottles in India, they decided to "pull out the SIM card in a bottle we were tracking, because we were using local SIM cards, and they put it in their phone and used it to text." This is the "tech density" that Microsoft's Brad Smith identified—the power and accessibility of technology are changing the way researchers work and the depth of the information that is available, as well as the quantity of data.

MAKING SCIENCE AND CITIZEN SCIENCE SYMBIOTIC

As they begin to rely on public data, scientists are improving protocols that increase the reliability. Studies demonstrate that thoughtfully designed citizen science projects can produce high-quality data. University of Michigan researchers found that users could accurately identify species in images from camera traps, which shoot photos when they detect motion, 97 percent of the time.[242] There are problems, however, with relying entirely on citizen scientists to identify species, even when the accuracy rate is fairly high.

First, it is difficult to find enough volunteers to work through the quantity of data created by camera traps. Although Zooniverse made it significantly easier for nonscientists to classify species and speed up the process of classification, it can still take years to get through a data set. While that reduces the burden on scientists, who can spend time on other work, they must still manage the data, respond to users' questions, and oversee the work.

The more interesting problem is one of bias—how humans and artificial intelligence make decisions when they are uncertain about a classification. The purpose of many of the camera traps is to identify and track rare animals. When faced with a difficult classification, choosing between a common animal and a rare species, humans have a bias toward choosing the rare animal. We want to see the rare species because it is more exciting. After all, that is why people volunteer to spend time looking at photos. As a result, humans overcount uncommon animals in camera trap photos.

Artificial intelligence, on the other hand, plays the odds. If an AI system has to decide between a species that shows up 95 percent of the time or the species that is present the other 5 percent of the time, it will err on the side of likelihood. Mathematically this is a safe bet. If the purpose of the project is to find endangered animals, this approach will undercount them, discarding photos and data that could be useful to scientists.

Improving artificial intelligence and computer vision can solve both of these problems. Sara Beery, a researcher at Cal Tech, launched the iWildcam competition to improve the ability of AI to identify species and overcome these biases.

Beery's route to becoming an expert in the world of wildlife and technology was circuitous. From the time she was young, her goal was to become a ballerina. "My whole life at the time was becoming a professional ballerina and [I] spent six years in ballet," she told me. Although it was what she wanted to do, it did not pay well. "One of my jobs was in Georgia near Georgia Tech," she explained. "I was pretty broke, so people would put posters up in neighborhoods for seminars because it was for college students, and they would offer free food." Looking to save money, she would attend the seminars for the food and would occasionally stay to hear the lectures. Soon, she became intrigued by the discussions of environmental technology and decided to make the leap from ballet to electrical engineering in the hope that she would learn enough to use the skills building environmental technology. Her studies took her ultimately to a PhD at Cal Tech and into the field of computer vision. She began with internships at Microsoft's AI for Earth program and Panthera, a project to locate snow leopards using camera traps.

Since she began working in the field, Sara has been trying to find solutions to the challenges of wildlife tracking using citizen science and camera traps. While at Microsoft, she helped develop what came to be called the "MegaDetector," an algorithm that can identify the presence of an animal but not the species. The site for developers looking to use the MegaDetector explains, "Conservation biologists invest a huge amount of time reviewing camera trap images and—even worse—a huge fraction of that time is spent reviewing images they aren't interested in. This primarily includes empty images, but for many projects, images of people and vehicles are also 'noise,' or at least need to be handled separately from animals."[243] Sorting out unneeded photos reduces the workload for scientists and, potentially, the citizen scientists helping them classify photos. "Machine learning can accelerate this process, letting biologists spend their time on the images that matter," notes Microsoft's developer site, by identifying photos that have animals, people, or vehicles. The remaining work of determining the species can be left to humans. "Machine learning is not so great at species ID, but we can detect animals, so we are removing the 70 percent of photos that are empty," says Beery. "It has been a really big success story, but it doesn't solve anyone's problem all the way."

The problem of correctly identifying species and then counting individual animals is much more difficult and is the focus of several iWildCam competitions. The 2020 competition was designed to improve computer vision to recognize species in a variety of locations. AI that is successfully trained to identify species at one location may find that a different background or set of animals makes identification difficult. Beery noted that if camera traps in the Serengeti were moved to another location in

Kenya, the identification system would struggle even with small differences in setting.

The 2021 competition built on that problem, asking competitors to determine how many individual species appeared in a series of photos. In order to ensure camera traps capture the correct species, "images are taken in motion-triggered bursts to increase the likelihood of capturing the animal(s) of interest." This creates difficulty when trying to come up with an accurate count of individuals. "For example," the problem statement for the competition explained, "if you get three images taken at one frame per second, and in the first you see three gazelles, in the second you see five gazelles, and in the last you see four gazelles, how many total gazelles have you seen?" This is more challenging than strictly detecting and categorizing species, as it requires reasoning and tracking of individuals across sparse temporal samples."[244]

Both competitions used camera trap images as well as images entered by citizen scientists into iNaturalist. The 2020 iWildCam competition specifically wanted to answer whether computer vision systems could "leverage data from other modalities, such as citizen science data and remote sensing data."[245] The competitions have anywhere from forty to three hundred teams competing, with the winners receiving a small financial prize. The knowledge and lessons from the competition are sent out to the community of researchers to refine the next generation of computer vision algorithms.

The success of the competitions and the improvements in computer vision are making the contributions of citizen scientists more significant. "I'm incredibly optimistic about the value of computer vision and machine learning," says Beery. The chance of computers replacing humans, however, is very small because while improving, AI simply cannot match the quality of trained experts or even volunteers. Beery says there is a danger in falling prey to "techsaviorism"—the belief that technology can come to the rescue and quickly solve difficult environmental challenges. The problem occurs when technology experts hope to apply a solution to a problem without also having some familiarity with the topic area. Ecologists work with the tech experts but find that after spending time and resources, the solution fits only a narrow circumstance or works well one year but not the next. Field researchers end up having to do much of the same work and become cynical about the application of technology to their field. Ultimately, technology complements but does not replace humans. Indeed, many of the iWildCam winners have a combination of experience with both technology and wildlife. Human judgment is needed, both to tune the artificial intelligence algorithms and to ground-truth the results.

Citizen science is going to become more important as time goes on. Beery explains that improvement is "contingent on platforms like iNaturalist. Those platforms are places where improvements in computer vision can be immediately translated." The more research-grade data provided by people around the world, the more useful data that is available to computer vision experts so they can calibrate their models. "That is going to be a huge thing in the coming years," says Beery, "the development of repositories, where data can be shared and pooled to train models."

Citizen participation is already having an impact in more than one field of research. As far back as 2014, Caren Cooper, an expert in citizen science, found that the study of the impact of climate change on migratory birds benefited from citizen science data. Cooper and her colleagues found that "between 24 and 77 percent of the references" for claims made in a study of the issue were backed by citizen science data without being specifically acknowledged.[246] The citizen science is there, and research is reliant upon it even if it isn't obvious. Another study of the use of citizen science in monarch butterfly research found that between 1940 and 2014, 17 percent used data from the public.[247] That number is likely to increase as technology makes participation easier.

That information is not just useful for scientists. Engaging the public to create usable information democratizes environmental research. Leopold, a scientist himself, lamented that "wildlife research started as a professional priestcraft." He argued, even in 1943, that this need not be the case. "The more difficult and laborious research problems must doubtless remain in professional hands, but there are plenty of problems suitable for all grades of amateurs," he said. "In the field of mechanical invention research has long since spread to amateurs. In the biological field the sport-value of amateur research is just beginning to be realized." Technology has closed the gap between scientist and interested amateur, changing the relationship from priest and parishioners to one of collaborators. As the impressive numbers in iNaturalist and the dramatic expansion of citizen science efforts around the world attest, Leopold's vision of engaging amateurs and making their contributions meaningful is being realized today.

MORE THAN SCIENCE

Sometimes connecting with nature simply means enjoying the majesty and beauty of the environment. Much of the discussion so far has focused on ways personal technology enables personal action to protect the environment. Underlying that is an appreciation for nature that creates the desire and interest to improve our stewardship of the planet. Simply getting out into nature is an important part of that. Immersing yourself in the outdoors does not just engage the mind, it is good for your health.

Several studies find that when people are in natural environments, they are happier and healthier. For example, one study by researchers at the London School of Economics found, "On average, study participants are significantly and substantially happier outdoors in all green or natural habitat types than they are in urban environments."[248] How did they find this out? With the assistance of smartphones (of course).

Using the GPS in a user's phone to determine location, a smartphone app called Mappiness would randomly ask questions of the more than 20,000 participants in the study. Users indicated how happy, relaxed, and awake they were, along with some additional information about who they were with and what they were doing. When someone was outdoors, the app also considered the weather. More than a million responses were provided, and the results were robust if not surprising. Study authors found that people were happiest outdoors, in marine and coastal environments rather than in an urban environment, with responses about six points higher on a 0–100 scale. "Alternatively expressed," they explained, "this is a difference of 0.28 standard deviations, or one of similar magnitude to, for instance, the difference between attending an exhibition and doing housework."

That is not the only study to confirm the power of nature to lift our mood. A 2021 study found that self-reported happiness among 26,000 people in twenty-six European countries was strongly correlated with the presence of a diversity of birds. "We found a relatively strong relationship," they reported, "indicating that the effect of bird species richness on life-satisfaction may be of similar magnitude to that of income."[249] Author Florence Williams agrees. Her book *The Nature Fix* examined the psychological and physiological benefits of nature. She wrote, "Psych studies using birdsong consistently show improvements in mood and mental alertness. An experiment at an elementary school in Liverpool found that students listening to birdsong were more attentive after lunch than students who didn't listen. Amsterdam's Schiphol airport plays birdsong in a relaxation lounge that also features fake trees. People love it."[250]

Although it is nice to enjoy the bird sounds at Schiphol, there are certainly better places to benefit from the sounds and experience of nature. There are many apps that can help you find good places to hike, but many are little more than online versions of books or maps. They do not take advantage of the connectivity offered by smartphones.

Smartphones and personal technology have changed the way people connect, breaking down barriers and creating opportunities that did not exist before. Campers, hikers, hunters, and fishers who want to find a special place—something unusual, inaccessible, or away from others—can now turn to several apps that help find exactly those experiences. Many are on private land, and simply getting permission to cross, let alone camp on, someone's land is not always straightforward. One challenge can be that while landowners are willing to allow hunters or hikers to cross their property, getting permission can be difficult. Knocking on doors can be awkward and landowners are not always at home. Now there are ways to address these problems.

"With the tap of your smartphone, you could rent hiking access to a local property, book a fishing pass to a neighbor's spring creek, or lease short-term hunting access on a nearby ranch," wrote environmental policy expert and avid hiker Shawn Regan of the Property and Environment Research Center.[251] Connecting people to conserve and enjoy the outdoors is the mission of Hipcamp, an app that is similar to Airbnb for camping. "By connecting people with the land and each other," their website explains, "Hipcamp works to support those who care for the land and get more people out under the stars. We do this because we believe humans in nature bring out the best of human nature."[252] App users can find places to camp, park an RV, or even rent a treehouse. The app also includes locations at more than 16,000 national, state, and regional campgrounds, but the real innovation is connecting campers with private landowners. "By connecting landowners who want to keep their land undeveloped with our community of responsible, nature-loving Hipcampers, recreation can help fund the conservation of this land."

Hipcamp isn't the only tool that can help people experience nature firsthand. Hunters can use apps like onX which show property boundaries to ensure they are not trespassing. While they are out there, they can also hunt for another sought-after item: morel mushrooms. Morels often appear on burned soil in the spring after a wildfire, and onX allows fungi hunters to show recent fire information on their map to help them find the sought-after item.[253]

The quality and variety of information available to nature lovers will continue to grow with time, making it easier and more enjoyable to get outdoors.

A STRONG CONNECTION CREATES ACCOUNTABILITY FOR OUR ACTIONS

A strong connection to the environment, whether that is citizen science or an appreciation for the beauty and majesty of nature, has another important benefit: it makes clear that we are accountable for the impact of our actions and need to take actions that are about more than public displays of righteousness. The more we understand the complexity of ecosystems and appreciate the intellect of animals, the more likely we are to make sure the things we do to protect them are more than window dressing.

There is a thirteen-second video of a happy coyote playfully waiting for a badger before they walk together through a culvert under a busy California highway. It has been viewed more than twenty million times, and the tweet accompanying the video notes, "Coyotes and badgers are known to hunt together."[254] The complexity and mystery represented by even this short glimpse at animal behavior add to our reverence for the natural world. The ability to collaborate is more common and sophisticated than we usually think it is among wildlife. Our human response to animal behavior we do not understand is to ascribe it to "instinct," which often simply means we don't know why animals do what they do. The case of the badger and coyote, however, is hard to explain this way. Collaboration requires communication that expresses both the intent to collaborate and a strategy. How do a random coyote and random badger, coming across each other, make it clear that they are not hostile but also that they want to work together? How do they build trust? How do they decide when to go hunting and where? How do they decide to share what they catch? Citizen science offers the opportunity for anyone to answer these types of questions while simultaneously creating a sense of wonder about the complexity of the world. It turns an abstract belief in environmental protection into an ethic of stewardship that encourages us to hold ourselves accountable by making sure the actions we take are effective.

The connection to nature that is stimulated by the coyote and badger video and citizen scientist questions affirms why we must be serious about environmental policy decisions. Making the right environmental choices, personally or in our public policy, is difficult. We cannot always know how best to help protect wildlife or to reduce our environmental impact. The respect we feel for wildlife and the recognition of common traits requires that we be humble and honest, acknowledging when we make policy mistakes and holding ourselves, and especially policy makers, to high standards of success in ways that truly benefit the environment, rather than just making us feel good about ourselves.

What's the Downside?

Many hands make light work, and small personal technologies are enhancing the power of each environmental contribution. As powerful as this approach is (and it will continue to grow), it is not a panacea. The underlying philosophy of harnessing the power of small contributions is that incremental change is more likely to be effective and durable than hoping for revolutionary change directed from the top. That mindset acknowledges that these solutions take time and are incomplete. The approach also comes with new challenges. Some are already raising concerns about the ability of technology and individual action to tackle the scale of environmental challenges we face today. Others express concerns that there are security risks associated with reliance on internet-based technologies and IoT devices. There is also a concern about privacy.

And, of course, there is always the threat identified by John Inglis, the former deputy director of the National Security Agency: "I don't think paralysis [of the electrical grid] is more likely by cyberattack than by natural disaster. And frankly the number-one threat experienced to date by the US electrical grid is squirrels."[255]

These are legitimate (some more than others) concerns, and if individuals empowered with small technology are going to supplement existing public policy and solve problems that have confounded government solutions, they will need to be addressed. The good news is that many of the issues are either exaggerated or can be managed.

CAN SMALL BE BIG ENOUGH?

The environmental challenges we face in the twenty-first century come in a range of sizes, from pollution in a local stream to over-fishing in the oceans and the increase in ocean plastic. When it comes to big environmental problems, however, there is none matching the planet-wide scope of climate change. Whether it turns out to be an existential crisis or a problem that is real but manageable, reducing the risk is a challenge that involves and impacts everyone. It seems logical that the scope of solutions to such a wide-ranging problem must be equally vast. In many ways this makes sense. Rather than trying to engage billions of people around the world individually, it is much easier to work with governments and corporations whose decisions will affect large numbers of people.

For example, a report in 2021 claimed that just twenty companies were responsible for 55 percent of the world's single-use plastic waste. The companies, which included Exxon Mobil and China's Sinopec, were contributing to climate change because "nearly all the single-use plastic manufactured by these companies—98 percent according to the report—is made from 'virgin' (fossil-fuel-based) feedstocks rather than recycled materials."[256] If governments or public pressure could reduce the climate contribution of just a few companies, it might have an enormous impact. This is the issue of transaction costs—it simply costs too much to try and convince every consumer to stop buying plastics made from fossil fuels. The theory is that focusing on a few big producers is simpler and holds the potential of major advances in a short period of time. It is also easier politically to blame corporations for pollution than consumers (and constituents) who like those products.

Even in articles that purport to highlight personal action, the emphasis is often on influencing the political process. In an article titled "5 ways you can personally fight the climate crisis," the World Economic Forum suggests that people use a "climate lens" in the daily decisions we make about what we eat, where to shop, what to buy, and where to work.[257] The article does not explain how to do that practically, emphasizing instead that such decisions "start the discussion" and can influence others around you. Rather than highlighting tangible ways to personally reduce environmental harm, the focus is on the impact our behavior has on a larger social movement. The other four ways they list are all political, including using relationships to "influence the masses"; learning about the "local, regional, national, and global policy landscape"; and telling people to "amplify the voices of others." The assumption in each of these is that only political efforts can yield the changes necessary to meaningfully reduce CO_2 emissions and address climate change.

Others are more explicit in their belief that small, individual-based efforts are inadequate. British climate activist George Monbiot dismisses actions that are less than systemic, calling them "pathetic micro-consumerist bollocks."[258] Other than becoming vegetarian, he argued, individuals are simply not capable of making the magnitude of change that is necessary. Focusing on small solutions, he worries, is a distraction from putting energy and emphasis on political action, where the real solutions lie.

While I understand the frustration and impatience that motivate these sentiments, they have been the guiding mindset behind the environmental failures of the recent past. As seductive as big, systemic changes are, they can offer a false hope. Many politicians promote this mindset because they believe it but also because it magnifies their own significance. People run for office because they want to do meaningful work, and tackling the world's largest environmental challenge is certainly meaningful. There are few incentives for elected officials to downplay their role with voters, so they put themselves at the center of the solution, even when that isn't appropriate.

In many cases, political solutions are simply not up to the task of delivering environmental results. For climate change, there are many examples of these shortcomings. In some instances, voters changed direction because of the financial cost of climate policy. Australia,[259] Ontario,[260] and Switzerland[261] all rejected or overturned government climate policies in recent years. Elected officials frequently fall short of their promises. Many of the countries that signed the Kyoto Protocol to cut CO_2 emissions missed their 2012 target. In 2021, the Climate Action Tracker, which compares government actions to the pledges of the Paris Climate Accord, found very few countries were taking action consistent with those promises, including the European Union, Canada, Brazil, Australia, Japan, and China.[262] The bigger the problem, the more difficult it is to coordinate political efforts. The assumption that political solutions will be effective or sustainable once adopted is contradicted by recent experience.

Large solutions are seductive because they are visible. Visibility, however, does not always equate to results. By contrast, it is difficult to perceive the many small actions that add up to meaningful change. Charles Darwin provided a good example of this phenomena in *The Origin of Species*.[263] He hypothesized that the number of house cats influences the amount of red clover near English villages. The clover relies on bumblebees for pollination, but since mice destroy the bees' in-ground hives, the more mice there are, the fewer bees. By eliminating mice, house cats help protect bees, which means more pollination and more red clover. Understanding the many small influences leading to the presence of more red clover is extremely difficult, but small changes can yield large and apparently unconnected results. The evolutionary nature of incremental changes can obscure the power of those forces.

The ability of technology to multiply the effect of many small efforts should already be apparent. Citizen scientists overwhelmed Galaxy Zoo in the first days after it was announced. iNaturalist now has one-sixth of all species in its database. BirdNet, an app to identify bird sounds, had more than 50 million observations, with nearly 80,000 users a day in 2021.[264] Plastic Bank, about which we will learn more, has already stopped a billion plastic bottles from reaching the ocean, and Seabin has already collected more than four million pounds of trash that did end up in the ocean. Homes with a Google Nest have saved more than seventy billion kilowatt hours of electricity, and car sharing reduced the number of vehicles in just a handful of cities by nearly 30,000. Curtis Tongue, the cofounder of OhmConnect, a company that uses smart thermostats and smart plugs to help people save energy, estimates the potential electricity saving using these simple tools in the US alone is equivalent to the generation from thirty Grand Coulee dams—the largest electricity generator in the United States.[265]

These are just a few examples, but they are evidence that technology-based environmental empowerment can be a powerful force for the planet.

This is just the beginning. The history of innovation reveals that in the early years, the promise of technology is greater than the actual performance. As time goes on, technologies mature and innovators use them in ways that were unanticipated, and the effect surpasses even early expectations. Science writer Matt Ridley describes the process this way: "Along comes an invention or a discovery and soon we are wildly excited about the imminent possibilities that it opens up for flying to the stars or tuning our children's piano-playing genes. Then, about ten years go by and nothing much seems to happen. Soon the 'whatever happened to...' cynics are starting to say the whole thing was hype and we've been duped. Which turns out to be just the inflexion point when the technology turns ubiquitous and disruptive."[266] He calls this cycle "Amara's Law," for Stanford University computer scientist

Roy Amara, who first identified the pattern of hype, disappointment, and breakthrough in many innovations. "Amara's Law implies that between the early disappointment and the later underestimate," Ridley writes, "there must be a moment when we get it about right; I reckon these days it is fifteen years down the line. We expect too much of an innovation in the first ten years and too little in the first twenty, but get it about right at fifteen." Smartphones themselves—let alone environmental applications—are not even fifteen years old, so we are still early in this process. Given the promising results we've already seen and the steady growth of small technology, there is reason to believe Amara's Law is at work for environmental technology as well.

Projections of the growth of small technologies and Internet of Things (IoT) devices back this up. One estimate shows the total number of IoT devices for all purposes (not just for energy and the environment) nearly tripling between 2020 and 2030 to more than twenty-five billion devices.[267] That estimate looks tame compared to the projection from the US Department of Energy's Pacific Northwest National Lab, which predicted eighty billion devices by 2025.[268] In the energy efficiency sector, that translates to connected water heaters, EV chargers, heat pumps, smart thermostats, and many other devices that help users become more energy efficient and save money.

It is important to remember that even if these technologies do not "solve" the problem of climate change or ocean plastic or anything else by themselves, they are making a growing contribution. Many of the environmental problems we find challenging to address are much smaller in scope. The pop-up wetlands created in California using eBird to help migratory birds, or the water provided by eWATERservices pumps in African villages—both make the lives of people and animals better.

There is a story that has been adapted in many ways about a young girl walking along a beach where thousands of starfish have been washed up and are dying in the hot sun. As the girl picks them up and throws them into the water, one person who is watching tells her, "There must be tens of thousands of starfish on this beach. I'm afraid you won't really be able to make much of a difference." The girl thinks, and then picks up another starfish, throws it into the ocean, and says, "It made a difference to that one." We should not ignore the magnitude of some environmental challenges, but belittling the thousands of small efforts that make a difference in their own way is counterproductive.

THE RISKS OF CYBERSECURITY

Another concern is the increased risk of cyberattacks as the number of small technologies—environmental and otherwise—increases. Connected devices that are part of the IoT are effective because they access and share information rapidly and respond quickly. That capacity can be a weakness. The risks can be found even in the most mundane of places.

Smart lightbulbs, which can be controlled from a smartphone and set to turn on and off when users leave the house or on a timer, can be hacked. Experts found security weaknesses with Hue lightbulbs, allowing someone to take control of the bulbs, changing their color and brightness.[269] When users try to reset the bulb, hackers could install malware to take control of an entire network. The weakness was identified not by actual hackers but by security experts trying to discover potential vulnerabilities. Actually taking over lightbulbs in this way would be difficult—a drone hovering over the location was used in one of the tests. The fact that even smart lightbulbs can be hacked does demonstrate, however, that as more of our appliances become connected, we have to pay attention to the cybersecurity risks they will pose.

There have already been some significant demonstrations of the potential threat. In 2016, hackers took control of millions of IoT devices and launched an attack that shut down several web pages, including Twitter and Spotify, and caused widespread outages in the US and Europe. Called a "botnet" attack, it used connected devices with inadequate security to send a massive stream of requests to service providers on the internet, overwhelming their ability to process them. Known as a "distributed denial of service" attack, it was possible because of the large number of devices the hackers controlled. After the attack, security expert Bruce Schneier explained the strategy. "The attacker can build a giant data cannon, but that's expensive," he wrote in *Security Intelligence.* "It is much smarter to recruit millions of innocent computers on the internet." The devices included webcams, wireless routers, and video recorders, many of which had not changed their log-in and password from the factory setting. "On large networks, IoT devices are sometimes deployed as shiny new equipment but are then neglected, missing regular maintenance such as monitoring and updating firmware, and left with nothing but default passwords as a layer of protection from external intrusion," wrote Charles DeBeck, a cyberthreat expert at IBM.[270] The increasing number of IoT devices, which helps overcome the problem of scale when addressing large environmental problems, can also be a growing headache if they are not secured properly.

As the electrical grid becomes more distributed and interconnected, it also provides more avenues into hacking the system. "Five, ten years ago, you had a network that was not really instrumented with any digital devices, so it was all static and physical security was really all they're worried about," said Mike Kelly, senior research analyst at Guidehouse in *IoT World Today,* an online magazine for the IoT industry.[271] "But when you have billions of devices—whether on the power lines, at the substation, in the homes—you essentially have this entirely new network of devices that are vulnerable to attack."

Professor Stu Steiner of Eastern Washington University is part of a program to train students to detect attacks on small cities where they may not be able to afford a cybersecurity expert. One of the entities he monitored installed an EV charging station that was connected to the internet to share information about usage and availability. The organization didn't set up any security. "You could log into it and update the firmware," explained Steiner. Without the security, someone could take control of the charging station and "cycle the battery [so] that you could cause the car to explode," he told me. Recognizing and addressing those potential attacks requires skilled staff that many organizations simply don't have.

Although many of the attacks come from individuals or small groups, there is also the danger of state actors. Two grid outages in Ukraine have been blamed on Russia, and in the first meeting between President Biden and Russian president Putin, the American president provided a list of key infrastructures, including the energy sector, that must be considered off-limits or risk a counterattack by the United States.[272]

Achieving the promise of small environmental technologies will require robust security protections. If these systems are repeatedly used to attack the backbone of the internet or cause economic damage, people will be hesitant to use them, fearing the loss of privacy or control. The good news is that many of the threats can be stopped with some simple measures. An appreciation for the growing threat motivated the US Department of Energy and the National Laboratory system to step up their efforts to identify potential threats and invent techniques to stop hackers.

Some solutions to this complex problem are fairly simple. The 2016 botnet attack took advantage of the fact that many people did not change the default password on devices. Hackers compiled a list of standard log-in and password information and tried those combinations on the devices they found. That shotgun approach allowed hackers to take control of millions of devices. Steiner explains that simply getting people to take cybersecurity seriously will make a big difference. Using an analogy from the COVID pandemic, he says, "If we get to 70 percent of people being aware, we can get herd immunity."[273] One step he hopes will become common is for companies to create unique passwords for each IoT device. He believes that small changes like that, along with more attention to the problem, will reduce the overall risk.

But that alone won't solve the problem. There is a growing need for cybersecurity experts to maintain systems, updating them against new threats. Some estimates indicate there will be nearly two million unfilled cybersecurity careers in the near future.[274] Filling those positions is an important part of fulfilling the potential of environmental IoT devices. Steiner's program at Eastern Washington University joined a group of state schools training students to recognize potential intrusions. The program, called the Public Infrastructure Security Cyber Education System (PISCES), has students monitor the traffic for small cities and report anything that looks suspicious.[275] The Department of Energy is also helping with innovative programs like the CyberForce Competition, which lets college students learn about hacking threats while competing with others. Programs like these create the expertise necessary to stop future threats while maintaining the usability of IoT devices.

Stopping future hackers will also require more research to identify weaknesses before they are exploited. The Pacific Northwest National Lab (PNNL) is located in Richland, Washington, next to the site where the nuclear material was created for the first atomic bomb test. PNNL created the IoT Common Operating Environment to study cybersecurity risks associated with adding more connected devices to the energy grid.[276] The program provides a test bed for dozens of IoT devices, allowing researchers to discover vulnerabilities and design countermeasures. The goal is to maintain the functionality of energy-saving devices while improving security. PNNL built replicas of smart homes, testing combinations of IoT devices and their vulnerability to attack.[277] The test bed is managed by Penny McKenzie, who recognized the need to create an environment that could test and monitor the risk of cyberattacks. Her concern for the environment is evidenced by the two bumblebee tattoos she proudly displays, telling me that she designed her garden to make it hospitable for the fuzzy pollinators. She is also excited about the promise that environmental technologies offer. "As I started going through and looking at the operational environmental technology environment, I noticed that IoT was in everything," she told me. A key element of her work is to simply understand what normal behavior looks like for devices. By setting a baseline, it becomes possible to detect when something is wrong. "The challenge right now is the security, and it has a lot to do with there [being] so many different vendors and manufacturers," she said. "There is no one way a device does something. There is no universal language when it comes to IoT." Although she jokes that there will always be a need for her job, she believes the threats can be managed in ways that ensure IoT devices are still usable. "There is a way of addressing the problems without going crazy policy-wise."

In one way, distributing electricity generation makes it more difficult to have a large impact on the energy grid. Diversity creates the flexibility to adjust if one part of the system is shut down. Large generators, like a hydroelectric dam, are more likely to have cybersecurity experts protecting their systems, but they are also high-value targets. It is more difficult to protect millions of individual connected devices, but it is also more difficult to shut down an entire system. Ronald Melton of PNNL explained, "While distributed generation and storage resources increase complexity from a grid management perspective, they also provide grid operators with options to maintain electric service even if a disruption occurs to a bulk power station." Distributed devices create more opportunities for attack but make each attack more manageable.

Despite these risks, Steiner is optimistic that we can manage cyberattacks. "I am more hopeful that we will get people aware, and the cybersecurity threats will be less," he says.

MISUSING THE POWER

The goal of creating small, personal technologies is to empower people to act on environmental threats. Some worry that the power provided to individuals can be misappropriated by government or others who can override personal decisions. The most dramatic example of the potential for misuse by politicians is the "social credit" system in China. Writing in Princeton University's *Journal of Public and International Affairs,* Eunsun Cho notes that the system was initially intended to "function as a nationwide incentive mechanism, by collecting social credit information from every individual and enterprise to reward trust-keeping and punish trust-breaking behaviors."[278] Initially, the system was limited to the economy, but it was soon expanded to the political sphere. "Although the original goal of reducing transaction costs in the market did not change, the scope of the system as outlined in the 2014 Plan was no longer limited to the economy," Cho wrote. "It was also meant to promote a culture of honesty and sincerity in all corners of society, including the government, the judiciary, private enterprises, and social organizations." Reducing the transaction costs of control makes inappropriate control more feasible. The ubiquity and control associated with China's social credit system rightly make people nervous about the trade-off between privacy and the power provided by connected technologies. China is an extreme example, but there are more narrow examples in the United States of personal technologies being used in ways consumers did not expect.

In June 2021, some residents of Houston, Texas, found that their homes were much warmer than they should be given the thermostat setting. On a hot afternoon, one owner in Texas discovered his house was 78 degrees inside despite turning the thermostat down. As outdoor temperatures increased and electricity demand grew, the setting had been changed by a third party in order to reduce the demand on the Texas grid.[279] What the owner, and others, did not realize is that they had been enrolled in an energy-saving program. In exchange for giving a third-party control of a smart thermostat when electricity demand is high, users are enrolled in a sweepstakes. In the Texas case, some were caught off guard. The example raises concerns about adequate disclosure and the potential for connected devices to be controlled without users knowing. Similar concerns were raised when it was discovered that video from Ring doorbells was shared with police stations in the United States when there was an active investigation occurring in an area.[280] After criticism, Ring changed the policy to require police departments to make requests directly of users.[281]

These examples are far from the control being exercised in China, and the government there is using much more than simple personal technologies. It demonstrates that connected devices can be used in ways that people do not expect or support. This is a concern that I share, and as technology becomes more pervasive—for environmental and nonenvironmental applications—we need to become cognizant of privacy protections and limits on who can take control.

However, it is important not to exaggerate the danger of typical interventions. In many cases the alternative to the risk that companies or governments may take control of devices is no consumer control at all. Although I do not want a utility to control my thermostat without my permission when demand for electricity exceeds generation, I prefer that to brownouts where all electricity is cut off to my house and others in the area. Utilities and government agencies already have control of electricity distribution, and controlling my thermostat in emergency circumstances is less intrusive than the alternatives, even if it feels more personal. This doesn't mean it is optimal, but we need to keep that context in mind when we make decisions about the trade-offs between using smart devices or systems that provide more privacy but less control.

There is no silver bullet solution to this problem, and it will evolve over time. Competition among a diversity of devices will help. Stories about the way companies use data will put reputational pressure to improve transparency and privacy protections as happened with Ring. Ultimately, the decision to use these tools will be—or should be—made by individuals.

The misuse of personal data from devices isn't the only potential risk to privacy. Other conservation technologies including camera traps and drones have been used inappropriately, creating unintended (and intended) consequences that harm people and conservation efforts.

Camera traps and drones are increasingly used to prevent poaching of animals as well as for research on endangered wildlife. These tools can be extremely helpful to multiply the effectiveness of conservation efforts by reducing the need for staff to be in many places. The cameras are often put where people may travel legitimately, creating potential for conflict. The ethics of this inadvertent surveillance is becoming more widely discussed within the conservation community. Researchers have identified several potential threats to be addressed.

One concern is that monitoring that begins with conservation efforts can suffer from surveillance creep, with local authorities using the data for political purposes. Writing in the *Conservation Science and Practice* journal, several researchers warned, "Conservationists deploying CSTs [conservation surveillance technologies] should also be aware of the risk of surveillance creep, where CSTs deployed for one purpose may end up being used for others."[282] Sometimes that misuse can be surreptitious. Other times it is overt. Trishant Simlai, a researcher at the University of Cambridge, found that authorities in some Indian communities were candid about using drones to watch and intimidate people, admitting, "Our basic mandate for using drones is to create 'psychological terror.'"[283] In another instance, camera traps captured a picture of a woman relieving herself and the photo was shared among local authorities.

Women know they are being watched and it changes their behavior and social interactions. One woman interviewed by Simlai noted, "We normally sing 'kumaoni' songs or talk loudly while collecting grass in the forest to keep away elephants and tigers, when we see the cameras we remain quiet, you never know who's listening." Gathering grass and firewood is also a social outlet for women in the village, and the presence of cameras or listening devices limits their conversations. Simlai witnessed a conversation where one woman began to talk about her husband, saying, "The other day he hit me," but was quickly hushed by the other women who warned, "Quiet! There is a camera attached here."

These incidents not only violate the purpose of the technology and harm human dignity, but they also create resentment toward legitimate conservation efforts. There are numerous instances where local residents have destroyed camera traps fearing how they might be used. Conservation technology groups like WILDLABS and researchers have begun to create a set of ethical guidelines to balance collection of data and the need for privacy. They recommend working with communities before surveillance technologies are deployed to address potential concerns. Using the least-intrusive technology necessary to achieve a goal is also recommended, as is transparency, including providing contact information about who controls the cameras. Not all of these recommendations are appropriate, especially when cameras are intended to prevent poaching and secrecy is important. Each circumstance will be unique and require applying different rules. Awareness of the risk, however, can help avoid conflict and improve the quality of the conservation data. "We are genuinely excited about the prospects for CSTs and the insights gained from their unparalleled data generating capabilities to advance conservation practice and research," several conservation researchers explained in *Conservation Science and Practice*. "Conservation efforts worldwide will ultimately benefit from more informed and ethical use of conservation surveillance technologies, both in times of crisis and otherwise."

PROTECTING SENSITIVE DATA

Privacy is not just a concern for people. The power of citizen science lies in the ability to increase the quality and quantity of information about endangered species. That information can be misused, revealing the location of animals we would rather remain elusive. The Cornell Lab of Ornithology has already taken steps to hide eBird data that reveal the location of species at risk.

The university program's protocol for displaying birds and entering them in the database has been tailored to address the danger presented by specific and timely location data. "For example, the Critically Endangered Helmeted Hornbill is hunted and killed so that its bill can be carved like ivory," the Cornell Lab explained on the eBird website.[284] "Parrots have long been exploited for the pet trade, with a couple species (e.g., Glaucous Macaw) already extinct, or nearly so, as a result of overexploitation." Rather than discouraging users from reporting sightings of rare birds, the Cornell Lab actually encourages people to continue reporting. The information from sightings can help researchers and others working to protect the birds. In their guidelines on reporting sensitive species, the staff at Cornell Labs write, "Sensitive Species listings allow eBirders to submit specific location data for at-risk species in a way that supports science and conservation, without risk of causing harm to the birds."[285] To preserve the value of the citizen science information provided using the app, eBird limits how sensitive species are displayed. Sensitive species are hidden or have important data removed from public viewing. For example, the location of a sighting is reported with reduced precision to within four hundred square kilometers. Data about the number of sightings is shared but not specific information, and they can be viewed only by the owners of the checklist that entered the sighting. The Cornell Lab does have some additional advice to protect threatened species. "We ask eBirders observing any of the species on our Sensitive Species List to use discretion in sharing sightings via other public platforms (e.g., Facebook, WhatsApp, webpages, listservs, etc.)," they write. "Revealing site-level records exposes the birds to risk from professional bird trappers, hunters, and/or pressure by birdwatchers and photographers and could cause significant harm to the conservation of these species."

iNaturalist has a similar approach, providing several levels of protection for species that are threatened or at risk from poaching. Users can mark their observations "private," meaning nobody—

including iNaturalist staff—can see that data. If you see a threatened species and want to keep a record for yourself but don't want anyone else to know, you can do that. If someone does share the location of a protected species, iNaturalist will use national and international lists, like the IUCN Red List, to hide the sighting or obscure it. They call it "taxon geoprivacy"—reducing the resolution of sensitive sightings by showing a large box where the species was sighted, with the actual location at a random spot within that box (not the center). Members of the iNat community can also flag a species and they will have a discussion about concealing the information. iNaturalist cofounder Scott Loarie says they want the iNat community to take ownership and responsibility for the quality of their information, which also means helping make certain the data are not used to harm species. He told me that iNaturalist is beginning to work more closely with governments and species protection organizations to make sure they are being responsible. "We need to make sure we don't fall prey to all the things that social networks do if you don't tend your garden," says Loarie.

The internet makes it easier to share information collected by citizen scientists and for poachers to buy and sell species that have been illegally captured or killed. There is now an organization that is putting the power of citizen science to work to fight traffickers. The Coalition to End Wildlife Trafficking Online was created in partnership with dozens of technology companies like Facebook, Alibaba, and Rakuten to detect and remove trafficked animals from social media and other online platforms. Citizen science is an important part of their effort. The coalition trains "Cyber Spotters"—volunteers who report suspicious listings for sale online and report them. WWF, one of the sponsors of the coalition, reported that citizen scientists have an accuracy rate of 94 percent when identifying questionable listings, preventing more than seven thousand potential sales by 2021.[286]

No single tool will stop the misuse of citizen science data. Newton's Third Law seems to be at work here. For every increase in capacity to help protect species provided by small technologies, there is an opposite reaction by those who want to misuse that capacity. The hope is that, contra Newton, the opposite reaction is not equal, and the magnified power of thousands—even millions—of citizen scientists can overwhelm the efforts of the bad guys.

ASSESSING THE TRADE-OFFS

Ultimately, the message of this book is how personal technology expands the options to address environmental threats and use of resources like energy and water. Like traditional, top-down approaches, this strategy comes with trade-offs.

Small technologies aren't going to solve every environmental problem. While the risks they present—to privacy, cybersecurity, etc.—are sometimes exaggerated, they are real. The right tools should be matched to the problem. In some developing countries, relying on political solutions to be effective simply isn't an option. In developed countries, they solve problems that elude the grasp of government agencies.

Having worked in and around government environmental agencies for two decades, I am frustrated that they too frequently ignore their own shortcomings, limitations, and unintended consequences. For small technologies to live up to their promise, we can't make that same mistake and must be candid about the risks so we can address them.

Solving the Biggest Problems in the World

For many of the biggest environmental problems in the world the primary barrier to success is not a lack of technological fixes or a failure to comprehend the problem. If you want to reduce your CO_2 emissions, there are many options from which to choose, including buying an electric vehicle or purchasing renewable energy. Reducing the amount of plastic in the ocean is as simple as picking up plastic waste. Concerned about the declining population of pollinators or monarch butterflies? You can plant milkweed in your yard. This sounds simplistic, but in a fundamental way it is true. The real challenge is applying those solutions at the scale necessary to make a difference. It is making repeated, tangible improvements rather than hoping for a moon shot that may never occur.

While breakthroughs in esoteric technologies like fusion energy capture the imagination and promise massive improvements in environmental sustainability, they are often more notional than effective. At least once a month a friend will send me a story of some potential new innovation that promises to change "everything" about some environmental problem. Virtually all of them end up falling well short of the promise. For example, Washington State governor Jay Inslee, who has tried to cultivate an environmentally friendly image, predicted in 2008 that by 2011 "meaningful amounts of cellulosic ethanol are becoming available at service stations across the country."[287] He wasn't the only one who expected cellulosic ethanol to emerge. In 2007, Congress set annual requirements for a certain volume of cellulosic ethanol to replace transportation fuels during the subsequent fifteen years.[288] The targets required 250 million gallons of cellulosic ethanol be blended in 2011, increasing to 8.5 billion gallons in 2019.

Cellulosic ethanol is made using agricultural and forestry waste, like corn stover and low-value tree limbs left over from timber harvests. It promised to be a plentiful and low-cost form of renewable fuel. A decade and a half after that prediction and the requirements of the farm bill, cellulosic ethanol was still largely unavailable. A court threw out the 2011 target for cellulosic ethanol because it simply wasn't available. By 2019, rather than the 8.5 billion gallons originally required, the EPA cut the requirement to 418 million gallons, about 5 percent of the original target.[289] Predicting the future is hard, and focusing on a few big technologies makes the failure of those predictions even more costly. The incentives to be "bold" and put many eggs in one basket is strong, even though it is risky.

What makes small, personal technologies so effective as environmental tools is usability, not sophistication. Imagine an electric vehicle charger that recharges a battery virtually immediately. That remarkable discovery is only useful if people know how to find it and use it. An app showing the location and availability of the charger, while far more mundane, would be nearly as valuable (if not moreso) as the charging station itself. The information provided by the app makes the remarkable technology usable and unlocks its potential. Technologies that appear mundane can make complex innovation more available, usable, and powerful. You can have the coolest technology in the world, but if it is not practical, it will be outperformed by small, simple technologies that overcome barriers to solving distributed environmental problems.

The real-world value of these principles can be understood by looking at how they are being applied to confront two of the world's most challenging environmental problems—the increasing amount of plastic in the ocean and decarbonizing electrical grids.

MAKING BANK WHILE REDUCING OCEAN PLASTIC

Each year, about 8 million metric tons (about 17 billion pounds) of plastic enters the world's oceans.[290] About 80 percent of that amount comes from developing countries where the systems of trash collection are poor, with plastic trash washing into rivers and, ultimately, the ocean. The quantity of plastic going into the ocean is so vast that one report estimated there will be more plastic in the ocean than fish (by weight) in 2050.[291] Concern about this trend has spawned several efforts.

One such effort, The Ocean Cleanup, calls itself "the largest cleanup in history."[292] The goal of this ambitious project is to collect plastic and other trash that is aggregating in ocean gyres. The best known of these is the Great Pacific Garbage Patch, which actually consists of two patches where garbage is drawn into the center of a vortex created by ocean currents. The center of these vortices is fairly stable and trash becomes trapped, making ocean gyres seem like a good target to remove plastic. That's the theory behind The Ocean Cleanup's approach. They developed a large net designed to move more slowly than the current, allowing the water to carry the trash into it. That passive approach reduces the risk to fish and marine animals. Dragging a net to capture the plastic would ensnare fish and other animals, like a giant trawler. Recognizing that most ocean plastic is delivered by rivers, they also created a system using a barrier that directs floating debris toward a barge with a conveyor belt, gathering trash into containers for disposal.

A project called #TeamSeas piggybacked onto The Ocean Cleanup's efforts, raising $30 million to help them and other organizations remove 30 million pounds of plastic trash from oceans, rivers, and beaches.[293] They promised to create "one of the biggest, baddest, most-impactful cleanup projects of all time."

After being in place for about a year, these approaches had removed a combined total of 464,920 kilograms (about 1 million pounds) of trash from the ocean and rivers.[294] That's a pretty good start, but it is a very small percentage of the total plastic trash that enters the ocean each year.

Despite the aspiration, there are some who question this approach. Marine biologist Jan van Franeker, quoted in *Science* magazine, said, "Focusing clean-up at those gyres, in the opinion of the scientific community, is a waste of effort."[295] He argued that it was more important to stop the input of trash into the ocean, noting that when the amount of industrial plastic entering the North Sea was reduced, research found that seabirds ingested 75 percent less plastic. The interview with van Franeker included this almost parenthetical note: "Critics also worry that the high-tech clean-up project could distract from less glamorous efforts" to reduce the amount of plastic that reaches the ocean. One such "less glamorous" effort requires only a smartphone and a scale to help reduce the amount of plastic reaching the ocean. And it collects thirty-four times more ocean-bound plastic.

Much of the plastic that ends up in the ocean starts as discarded trash that washes into streams and rivers. Stemming that flow, rather than collecting it after it reaches the ocean, is the goal of Plastic Bank, a business-oriented nonprofit that pays people to collect plastic trash in countries that are the source of most of the waste in the ocean. With locations in places like the Philippines, Haiti, Brazil, and Egypt, Plastic Bank sets up collection centers and pays for plastic gathered by members of the community. The plastic is then recycled and sold to companies like SC Johnson and Henkel, who use it in their bottles rather than buying new plastic made from fossil fuels. The system keeps plastic from the ocean, provides income for people in poor communities, and offers a source of plastic that reduces environmental impact. And it is run almost entirely from smartphones.

As Shaun Frankson, one of Plastic Bank's cofounders, explains it, "Anyone with a cell phone and a scale can run our program."[296] Frankson and the others at Plastic Bank recognized that collecting trash was not a job most people wanted and, in many cases, carried a stigma. The key to their program isn't merely that people get paid for the plastic they return, but that the job helps improve their life. They call the plastic waste "social plastic" because it is good for people and the community. "What if the use of plastic brings dignity to recycling?" Frankson asks. "It should be something that stops going into the ocean, but also an authentic plastic source. When people pick up a bottle, it should be social plastic." Many of the people they work with don't have access to a bank account, so Plastic Bank pays them through their app, giving collectors savings they can access. When people turn in plastic at one of the collection points, the plastic is weighed and sorted, and collectors are paid in a type of crypto currency that can be used for purchases from Plastic Bank. This makes the system extremely simple and portable. It also helps avoid some of the pitfalls of working where government structures are weak or corrupt. "They can bring plastic to one of our banks and get food, fuel, health insurance," said Frankson. "And they can deposit the value and really find the things they need." As part of their commitment to use plastic waste to improve the community, they also have a recycling program that allows parents to pay school tuition for their children.

As important as the technology is to engage and reward the people collecting the plastic, it is critical to adding value to the plastic that is collected—helping to verify the environmental benefits. On their web page, Plastic Bank explains, "Our app uses cutting-edge technology, such as Smart Contracts, to ensure that all transactions are secure and transparent."[297] That transparency is important to corporate partners. There is a sticker on each Windex bottle that declares it is made from "100-percent Ocean Bound

Plastic." With so much attention focused on companies that "green-wash"—making environmental claims that turn out not to be true or are exaggerated—the ability to show exactly where the plastic came from is important, both to live up to their own promise and to avoid criticism. Plastic Bank makes it possible for anyone to see how much social plastic is used by SC Johnson, and to audit Plastic Bank's network of suppliers, right down to the individual branch. On the "Impact" page of their website, Plastic Bank has a detail section that provides statistics about each branch, creating a "trust score" for every one of their partners.[298] It shows how many people bring plastic to a location, how much plastic has been collected in the past month, and the number of days since the last plastic sale. For example, on the Indonesian island of Lombok, people can check the statistics of any of the thirty-seven (as I write this) locations that collect plastic, including one near the town of Mataram that has twenty-two members and collected 7,357 kilograms in the past month. Some locations are better, others are worse. "The biggest thing with us is to be absolutely transparent and authentic," explained Frankson. "There are things we said no to because we don't want to be part of a marketing campaign. We want to have ongoing ethical supply chains." The smartphone-based system they created allows them to do that, both in how they pay people and how they can reliably track and demonstrate the environmental benefit they are providing.

That reliability and transparency are enhanced by the block-chain-based system that underlies their work. Each step of the process, from the initial plastic collection to the final purchase of plastic, is tracked to ensure the reliability of the plastic source and also that collectors get paid for the value of the plastic they provided. Frankson explains that if collectors are "not bringing in the right plastic, the branch won't buy it. If the branch hasn't fully sorted the plastic by type, by color, it's really hard to sell to the processor. And the processor really only buys the fully sorted plastic which then makes the highest quality."[299] For plastic to make its way from the beach to a bottle, each transaction must be verified. "We have a complete transaction record of that collection member to our branch," says Frankson. "We have a separate transaction record of that branch to our processor." With that evidence of the chain of transactions, purchasers like SC Johnson can be certain that they are getting what they paid for, and their environmental claims are defensible and accurate. "They want proof beyond proof, upon proof that any claim they made is a valid, legitimate claim." The blockchain allows that. The sophistication and transparency of blockchain also makes it suited to parts of the world where banking or other financial systems may be unreliable. Plastic Bank could be certain that the right people were paid.

This simple, small approach is generating big results. In 2020, Plastic Bank's collectors amassed more than 17 million kilograms (38 million pounds) of ocean-bound plastic—thirty-four times as much as The Ocean Cleanup. How is it that a program that relies on simple technology was thirty-four times as effective at reducing ocean plastic as a high-profile, high-tech program claiming to be the "largest cleanup in history"? To be fair, Plastic Bank has been around for a few years and had more time to build its program. Still, Plastic Bank has been remarkably successful with few resources because it employs many of the advantages offered by small technologies. Working with limited budgets in places where there aren't many options, they made a virtue of necessity. Plastic Bank's founders did not have the luxury of being seduced by big technology and grand scale, and that has worked to their advantage.

Fundamentally, their approach provides an incentive to recycle. Turning plastic waste into a commodity that can be banked to become "social plastic" is the foundation of their work. "When someone is in a state of poverty and just trying to survive it is impossible to think of the oceans and the community," said Frankson. Collectors get paid and are given value that is more than just a little money; it's an opportunity to improve themselves, their children, and their community. "Some of the daughters of the best collectors earned free scholarships," said Frankson. "We do career training, so they are not full-time collectors but use it as something they do."

It also successfully aggregates many small actions into big environmental results. The cost of participation in Plastic Bank's program is very low. People only need a cell phone to become a collector or a branch, where plastic is gathered for resale. The transaction costs of participation are extremely small. The biggest barrier to being paid to keep plastic out of the ocean and have it turned into a recycled product by a business on the other side of the globe is probably the act of bending over to pick up the trash. The result is that millions of pounds of plastic trash are diverted from their path toward the ocean. And while some of that plastic might be collected by projects like The Ocean Cleanup, much of the plastic that enters the ocean sinks to the bottom. By keeping the plastic from reaching the water in the first place, Plastic Bank's program helps reduce the impact of hard-to-reach ocean plastic.

I don't mean to pit these efforts against each other. Both play important roles, and The Ocean Cleanup is attempting to address a problem that Plastic Bank simply cannot reach. At first glance, however, many would dismiss the Plastic Bank approach, saying it is inadequate on the assumption the scale of the problem requires the grand approach of The Ocean Cleanup. Instead, simple technology that engages millions of small actions is generating a more significant environmental benefit than the traditional, high-profile approach of The Ocean Cleanup or #TeamSeas.

Plastic Bank's effort is also likely to be durable. Their model connects businesses and consumers to their work and to their success. People want to help reduce the amount of plastic in the ocean, and they can be certain that buying a product that uses Plastic Bank's material helps achieve that goal. Frankson calls it "empowered awareness"—providing certainty that the environmental benefits they promise are real. In addition to a label that shows consumers a bottle was made with "Ocean Bound Plastic," Frankson says they hope to have a system where people can scan a code on the bottle and discover the story behind where the plastic came from. Much of that is already available to Plastic Bank's partners. On Plastic Bank's web page, companies—and those wanting to verify their claims—can see how many plastic bottles have been collected, the total weight, the number of collectors, and the number of collection points. Companies can acquire recycled plastic from many sources at a relatively low cost. The value Plastic Bank offers is a combination of the material worth plus the social and environmental benefits. Transparency and accountability increase the value of the environmental benefit by making it tangible. Similar to George Akerlof's insight about the price of used cars—that uncertainty about hidden problems with a vehicle reduces the amount people are willing to pay—questions about the origin of recycled plastic would reduce the value of purported environmental benefits to companies working with Plastic Bank. A robust and transparent blockchain system virtually eliminates that uncertainty, increasing the price companies are willing to pay because they can be certain the material they buy would likely have ended up in the ocean.

With billions of pounds of plastic entering the ocean each year, the work being done by Plastic Bank is just the tip of the iceberg at this point. Its performance, though, is impressive given the approach it is taking. The counterintuitive reality that they are significantly outperforming a larger and more high-profile effort is due in part to their ability to harness several of the advantages of small technology I have highlighted. Using their smartphone-based system, they provide an incentive to clean up the environment. They aggregate the small actions of tens of thousands of people, which resulted in the collection of more than one billion ocean-bound plastic bottles. And they connect companies and their consumers to environmental results by making the benefits of their work transparent and auditable. Whatever the ultimate solution to reducing the amount of plastic in the ocean, Plastic Bank will be a consequential part of it.

CONSERVING ELECTRICITY, MONEY, AND THE PLANET

There is no single strategy to reduce CO_2 emissions and the associated risk from climate change. Reducing emissions from electricity at a reasonable cost while ensuring it is reliably available is an important part of an effective climate strategy. Electrical generation is a significant source of worldwide CO_2 emissions.[300] Great Britain's National Grid ESO, which manages the transmission of electricity around the country, has pledged to be zero-carbon by 2025.[301] Although politicians tend to focus on adding more renewable sources of energy to meet goals like this, it is unlikely that construction alone can achieve the target. Significantly reducing the carbon intensity of electricity will be difficult without changing the way people use it. The traditional pattern of electricity use, where demand increases in the late afternoon and evening, virtually guarantees reliance on dispatchable generation—power plants that can be turned on and off as needed. Solar, although generally predictable, cannot be used to meet increased demand in the evening. Wind is less predictable and is most available at night. Rather than simply trying to build our way out of the problem, we can shift demand to use renewable energy when it is available. This reduces the need to pay for new generation, cuts CO_2 emissions, and keeps costs lower.

That philosophy was the motivating force behind the creation of UK utility Octopus Energy. Launched in 2016, they set a goal of affordably reducing CO_2 emissions. "It is all about renewable energy and trying to get consumers on to renewable energy," said Phil Steele, who Octopus Energy listed with the impressive title of "Future Technology Evangelist."[302] Octopus has seen impressive growth and reached 2.1 million households after just five years.[303] That is driven in part by high customer ratings. Octopus began with a focus on consumer-oriented technology. Steele told me, "We are very reactive to customers," and the technology they developed helps "to cut costs and make it far easier to support."

The innovative utility combines rates that encourage people to use electricity when it is inexpensive and has a lower carbon intensity, with technology that makes it easy for customers to see those costs and react. In 2017 they launched "Octopus Tracker," offering prices that followed the rates on the wholesale market, telling customers a day in advance what the price would be. The next year they added the "Agile Octopus" tariff, with very low costs at night and much higher costs during the peak hours in the early evening. Finally, they offered "Octopus Go" for EV users, which has extremely low electricity rates early in the morning (when people can charge their car's battery) and a flat rate the rest of the day. "We launched Agile Octopus to encourage households to shift their consumption from the daily peak energy demand period of four to seven p.m. when we're paying very high wholesale prices," said Steele. If they can buy less energy when wholesale prices are high, they pass the savings on to the customer. And the savings can be significant.

For example, at 3:30 in the afternoon, customers could be paying less than three pence for a kilowatt-hour (kWh) of electricity. A half-hour later that price might jump to seventeen pence per kWh. When 7:30 p.m. rolls around, those prices plunge down to eight or nine pence. Price changes during the rest of the day are less dramatic, but they can go from a penny to four pence in just thirty minutes. Prices can even become negative, when Octopus actually pays customers to use more electricity. This typically happens in the middle of the night, when demand is low and there is an excess of wind power. In December 2019, when Storm Atiyah blew in to the UK, the amount of wind energy on the grid doubled in twenty-four hours, enough to power about fifteen million homes in the UK. That electricity had to go somewhere. Rather than turning other generating stations off, which is expensive, Octopus decided to pay people to use the short-term surplus. Between 1:30 and 6:00 a.m. on December 8, 2019, Octopus paid customers about a penny per kWh. As Octopus staff wrote on their blog, "Overnight storms triggered the energy equivalent of an 'everything must go' sale."[304] If you owned an electric car, you hit a minijackpot. During the storm, the director of operations for National Grid ESO saw what Octopus was doing and tweeted out, "A big thank you to all those #EV drivers and smart cookies, including everyone on #OctopusAgile, who helped us balance the grid last night."[305]

The data shows the approach is working, helping shift demand and keep prices low. A study commissioned by Octopus called "A consumer-led shift to a low carbon future," claimed, "The First results from our Agile Octopus half-hourly time of use tariff shows consumers shifted electricity consumption out of peak periods by 28 percent."[306] The result, they claimed, was that the average consumer would save about forty-five pounds annually compared to a fixed-rate plan from Octopus. This is the company's own report, and everyone uses electricity differently, so we can take it with a grain of salt. But Octopus and a growing number of independent programmers are developing online tools so we don't have to take their word for it—anyone can see how the Octopus rates impact their costs and carbon intensity.

Although Octopus provides the hourly rates customers will pay a day ahead of time, they also encourage anyone who is interested to create new consumer-friendly tools that make adjusting individual demand less difficult. All rate information is made available by Octopus on an application programming interface, commonly called an API, that any programmer can query for information and then use in their own app. They even held a "Hack Day" to encourage people to use the API in creative ways. "We wanted to enable other tech innovators in the energy sector access to this incredible data," Octopus staff explained, "so we…cracked open our Agile Octopus API for public development."[307] The Hack Day and opening access to their API spawned several tools.

When Mick Wall started playing with the Octopus API, he only wanted the information for himself. His home already had solar panels, so he appreciated the value of using electricity when it was available. As an Octopus customer, he wanted to know more about pricing, but the detail he wanted was not available. "You could download small amounts of Agile tariff data for a single UK region from the Octopus website," he said, "but not my region." That's when he started his "journey of discovery."[308] Teaching himself several programming interfaces, he created a program that downloaded the data into a spreadsheet. Since he already managed the web page for his running club, he created energy-stats.uk and shared what he found. Soon, people began asking him to generate graphs, and over a Christmas break he added the ability for visitors to download the data. The initial simplicity of the site, providing daily electricity prices for regions of the UK, is what made it useful. Since then, the site has added comparisons between various electricity tariffs as well as price information for those selling electricity to Octopus, typically from a rooftop solar panel.

As easy as energy-stats.uk is to use, Kim Bauters wanted to make it even more simple. "People in my family found it was hard to track, so I decided to create an app."[309] Initially, he created an app called Octopus Watch for the Apple watch but has now added versions for the iPhone and Android. Unlike his other apps that use AI to create soap and gelato recipes, the programming for Octopus Watch was straightforward: take the data provided by the API and make it digestible. "All of these incentives don't necessarily translate into a user doing something," he said. The amount of data could be overwhelming. "It is like the Google problem of having too much data," he said, referencing the trend toward "big data," where individuals and businesses are inundated by information. Octopus Watch takes all that data and translates it into simple guidelines about when to turn on energy-intensive appliances like clothes dryers and dishwashers. Before the app, "we didn't use timers on any appliances," said Bauters, "but now we routinely use the app to find the best slots."

The work of those early pioneers is snowballing, inspiring new innovations. "Seeing what others were doing with the Octopus Energy API was also a huge inspiration to develop something," wrote Jacob Tammadge, a gamer and programmer who goes by the name Jakosaur. "Mick Wall (Energy Stats UK), SmarthoundUK (Octopus Watch) & Max Sanna (OctoWatchdog) have created websites, automated Twitter Agile pricing updates & apps that I continue to use daily." The initial motivation for creating his site, called OctoComparison, was that his parents switched to Octopus, and he wanted to see if they could save money by switching to the Agile tariff. "After experimenting with an API supplied by a video game company," he wrote in an email, "I was curious to see what I could develop using the energy data provided by Octopus Energy API." Once Tammadge began developing it, he found ways to improve the usability and said that he "lost count of how long I have spent developing OctoComparison," but he did receive donations from some who appreciated his work.

His web page helped people determine when they could save money by shifting their electricity usage slightly. "If people could push running the dishwasher from after dinner until later in the evening or early morning, they shifted that consumption to make some extra savings then," he told me in an email. As time went on and wholesale prices changed, the calculus of how to save money became more complicated and the value of OctoComparison increased. "Whether Agile is still cheaper is now probably very dependent on different factors which varies for each household," he explained. "A few examples of factors are solar, home battery system, [whether it] imports little electricity from the grid, or [homeowners] can easily avoid high Agile prices." For an average homeowner, navigating that complexity would have been impossible a decade ago. Now, the answers can be provided by accessing a web page developed by a hobbyist programmer who started out just wanting to help his parents save money. It is a convincing example of how energy conservation and access to renewable power have been democratized.

Another person who is convinced is Jan Rosenow, the European program director at the Regulatory Assistance Project, an independent energy think tank. "The main reason I chose Agile is that we installed a heat pump," he told me.[310] "The interest was to see if we could get benefit out of load shifting." Where does one of Europe's leading experts on energy demand turn for information about how to manage his personal energy use? "I rely on third-party apps," he answered. Prior to switching to Octopus Agile, Rosenow was paying about fourteen pence per kWh on average. After switching and using the information he found on these independently developed apps, he was paying about seven or eight pence per kWh.

With a customer base willing to change how and when they use electricity to meet supply, Octopus entered the renewable energy market knowing there were customers who would respond to price signals and purchase solar power during the day and wind power at night, when it is available.

When the UK dropped its existing subsidy for rooftop solar panels in 2019, known as the "feed-in tariff," Octopus helped fill the vacuum that was left behind. "We've ended up in a place where government subsidy was growing the market but then there was nothing available other than a guarantee that people should pay something," said Phil Steele of Octopus. "It slowed the renewable market." The solar subsidy was replaced with the "Smart Export Guarantee," which required utilities to purchase energy from customers, competing for supply by offering different prices for the energy created from their solar panels.[311] With a customer base of motivated buyers and the right price incentives to turn around and sell that electricity, Octopus has consistently offered some of the best prices for rooftop solar.[312]

Octopus is going even further to take advantage of the ability to provide virtually instantaneous information about prices and the availability of renewable energy. In 2021, after purchasing two wind turbines in Yorkshire and the south of Wales, they launched their "Fan Club," an electricity rate that drops when the amount of wind energy increases. For members of the Octopus Fan Club in those areas, rates decline by 20 percent when their local turbine is spinning, and by 50 percent when it is very windy. Octopus tells members, "Check the rate you're paying online in realtime, and see forecast wind speeds to help plan your day. And with IFTTT [If This Then That] and SMS alerts, you can make sure you're getting the most from your Octopus wind turbine."[313] What was previously a complex and opaque system—with utilities generating electricity and homeowners buying whatever was available without paying attention to market prices or source of electricity—is being replaced with a system that provides consumers with information and options that are near utility-grade.

Of course, not everyone will use all of these options. But the low cost of information means that Octopus can offer niche products like this—just two wind turbines—and make it worthwhile. The barriers to experimentation, both for Octopus and the many individual programmers taking advantage of the information on their API, are so low that even very small efforts can pay off, increasing the diversity of options available to customers and the likelihood they will find an option that suits them. Much like the diversity of transportation options discussed in chapter 3, Octopus is providing a range of products to suit different customers. Those who are simply looking for low prices can find a rate structure that suits them and receive feedback showing how much they save. Climate-conscious customers can adjust their energy use to favor solar or wind power, even instantaneously, and charge their electric vehicle when renewable energy is plentiful and prices are low.

The key to Octopus's approach is the use of incentives and price signals. New tools like Nest, Ecobee, and Sense help homeowners conserve electricity. The Agile tariff is specifically designed to take advantage of personal technologies that make changing how we use electricity easy and rewarding. Those technologies are changing how utilities operate and even inspired the creation of at least one company. In 2018, Evolve Energy was built around the increasing connectedness of appliances. When I spoke to Evolve's cofounder Michael Lee in 2019, he explained that their model was to align the incentives of the company with consumers by charging a low, flat fee per month and working to save customers as much as possible on electricity costs each month. "We don't make money on volume. Whether we sell a lot or a little, we still make the ten dollars a month," he said. "A happy customer is saving money relative to other options. Our incentive is to help customers to use less during certain times. We effectively sell trust."[314] Small, personal technologies made that possible. "Smart thermostats are a tool that is very low cost and quite high-penetration in the market," said Lee. "Customers just want to lay in bed or sit on the couch and adjust the comfort of their home. We are able to tap into APIs that already exist to help them control and optimize for price and carbon.... Anything you buy today has an app for that item. If you buy a new appliance, it probably has an app. Even hot water heaters have an app. Anything that has an app, we can piggyback off that connectivity." Less than a year after I interviewed Lee, Octopus purchased Evolve, making him CEO of Octopus Energy USA. Just months after he became CEO of Octopus USA, Lee saw firsthand both the need to provide clear incentives to conserve electricity and the opportunity incentives plus technology create, not merely to help the environment but also to ensure electricity is available even during a crisis.

As the sun went down on Valentine's Day 2021, subfreezing temperatures across Texas sent demand for electricity in the state to an all-time winter high. Hours later, unable to keep up with demand, Texas grid managers ordered blackouts to prevent physical damage to the system. Much of the debate after the Texas blackouts focused on the failure of state officials to plan for an adequate supply of reliable energy. This planning failure was not unique to Texas. The Midcontinent Independent System Operator (MISO), which covers fifteen states, including part of Texas and the neighboring states of Arkansas, Louisiana, and Mississippi, hit an all-time high for demand in that region, which also resulted in officials ordering blackouts. Planning is valuable, but during a crisis it provides only a limited amount of flexibility.

Some pointed to the unreliability of wind power, because wind generation fell by half just as the cold front moved in. Others point to the failure of the natural gas system because some power generators were unable to receive enough fuel to meet the need. After the crisis, some argued that Texas and other grids across the country needed to implement more and better planning. Increasing reserves for extremely unlikely events, however, is expensive. Buying a backup generator for your home may be sensible; buying a second generator to back up the first one is probably needless. The same is true with excess electricity reserves.

Ultimately, any system that relies on planning for extreme weather events will be either very expensive or likely to fail at some point. Price incentives and technology create the ability to respond to a crisis in a dynamic way, adjusting in real time. The tools existed to do that in Texas, potentially avoiding what became a costly, and deadly, catastrophe. As the severe weather set in, electricity prices skyrocketed. In 2020, the average retail price of electricity for residential customers in Texas was just under 13 cents per kilowatt hour (kWh).[315] On the morning of February 15, the spot price was more than 150 times higher, exceeding twenty dollars per kWh. Those spot prices are eventually passed on to consumers in the former of higher average rates in the future. During the storm, however, consumers with low fixed rates had little incentive to reduce the amount of electricity they used. I spoke with Lee after the crisis and he said that people on fixed rates were driving demand higher, making it difficult for generation to keep up. Lee said there were some fixed-rate customers "who were using 500 kilowatt hours a day. Two weeks' worth of electricity in a single day." And why not? When the temperature is 12° F (−11° C) and energy is relatively cheap, the decision to keep the heat on is fairly obvious. Millions of people making that same decision contributed to the blackouts that cut power for everyone. ERCOT, the agency managing the Texas grid, was sending notifications to people asking them to conserve, but it made little, if any, difference. Demand kept climbing, until early in the morning of February 15, the system could not keep up and grid managers shut down some generation to prevent permanent damage as the grid became unstable. Even shaving a few percentage points off demand would have made a difference. Reducing demand by just 10 percent on February 14 would have kept it at a level the grid successfully met during each of the three days prior to the blackouts. Unfortunately, very few people in Texas had incentives to turn down the thermostat, so the system crashed.

Incentives also work the other way, increasing supply as well. Lee reported that as the price of electricity increased during the crisis, some customers with solar panels began conserving electricity so they could sell the excess energy for a big profit. "We actually had customers getting paid hundreds of dollars for the energy they were pushing out to the grid."

By making the incentives to conserve electricity transparent and providing the tools for customers to adapt, Octopus and the growing number of utilities like them are not merely fulfilling their mission to provide environmentally friendly energy. They are also empowering people, helping them save money and play an important role keeping the lights on when the weather doesn't cooperate.

HOW SMALL OVERCOMES OUR BIGGEST CHALLENGES

Plastic Bank probably won't solve the problem of ocean plastic. Octopus isn't going to end the risk of climate change. Both organizations are having an outsized impact, growing rapidly, and achieving more than traditional and eye-catching environmental approaches. Both have tapped into the power provided by democratizing environmental action. Small innovators create a virtuous cycle, where new technologies make it possible for individuals to act in a way that aligns their self-interest with positive environmental outcomes. Each project takes advantage of the principles outlined in the previous chapters.

Both efforts reduce the transaction cost of information and collaboration. By providing nearly instantaneous information about electricity prices at no cost, Octopus empowers people to take control of their energy costs. For virtually the entire history of electric power, balancing the supply and demand of electricity on the grid was only available to trained managers. Today, members of the Octopus Fan Club can do something similar by turning on their dryer or charging their EV when wind power is plentiful. Companies that buy material from Plastic Bank can be sure they are getting the environmental benefits they want because the system is transparent, making it extremely easy to get the information they need. This is possible because the costs of information and action are extremely low.

Armed with that information, people are rewarded for environmental stewardship. Octopus customers save money by using energy when it is cheap and carbon-free. Plastic Bank collectors are paid to keep plastic out of the ocean, and companies like SC Johnson can market their products to environmentally conscious customers. Relying on incentives rather than politics makes environmental actions more durable and more effective. Octopus customers who buy electricity when it is plentiful will save money no matter who is sitting in the prime minister's office. The governments in Egypt, Haiti, and Brazil are very different, one from another, but Plastic Bank's model works in all three countries because the system is built around individual incentives, not politics.

Reducing the cost of information and innovation also means efforts can be personalized and localized. Octopus offers a range of rate structures to suit different customers. They made their data available to encourage small innovators to create a diverse set of tools to help make use of those tariffs and keep customers happy. Understanding that they are working in places where government structures can be unreliable—or even corrupt—Plastic Bank offers several ways to pay their collectors, from a cryptocurrency to tuition payments, depending on the circumstances. There are very few circumstances where one size fits all. A diversity of options means more people can act in a way that is effective and suits them.

The cumulative effect of engaging those principles is that both organizations are having extraordinary success by taking the many small acts of conservation and restoration and multiplying them millions of times.

It is important to remember that not every environmental problem is as vast as these two. Small technologies can be scaled up, but are also suited to small, but important, environmental endeavors. People looking to eradicate invasive species in their area can use iNaturalist to identify and tag the location of noxious weeds for removal. The eBird database can be used to identify important local parcels of habitat. Rainforest Connection is working with groups as small as a few people, providing information about the location of protected species, helping ensure those areas have suitable habitat.

It makes sense that small technologies are capable of addressing localized problems. It is becoming apparent that they may also be the most effective tool at addressing the biggest environmental problems we face. And the full promise is only beginning to be realized.

Democratizing Environmentalism

During the first forty Earth Days, between 1970 and 2010, the approach to addressing environmental challenges, especially in the developed world, changed only marginally. The primary question was how to structure regulations and government spending to address environmental problems. Sometimes this worked. In a frustrating number of cases, it did not. For a growing number of innovators, that frustration turned into motivation. Whether it was surfers tired of plastic in the waves they enjoyed, biologists concerned about illegal trafficking of turtle eggs, or a tech entrepreneur who wanted to reduce water waste so she could enjoy her shower, the low cost of innovation became an opportunity to find new solutions and share them with others around the world who had the same concerns. Their goal wasn't to replace the successes we've already enjoyed from well-targeted government programs and regulations. They wanted to build on them, filling in gaps where top-down solutions were not working.

Small technology has democratized environmental action, shifting responsibility from a few people in official positions to millions who are working together and individually to make the world's ecosystems better and improve the state of our environment. Rather than outsourcing our concerns about the planet to politicians and government agencies, people can act. Looking for solutions that emerge organically from individuals and small groups is different than the model we typically associate with environmental action. Scott Loarie, one of the cofounders of iNaturalist, told me how counterintuitive a decentralized approach was to him when they first launched. "I came out of academia," he said. "I believed things were top-down," but he was open to learning how things worked in Silicon Valley. "iNaturalist learned from these and other efforts," he said. "I hope we are doing a good job of leveraging this with decentralized crowd-sourcing." That decentralization is a strength, not merely because iNaturalist and efforts like it take advantage of information provided by people across the planet. A decentralized approach gives people ownership of the results, so they are more invested in providing data that is accurate and usable, contributing to meaningful environmental action. "This is one of the things the environmental movement lacks," says Loarie. "Many environmental efforts are seen as these Big Brother things, and we need to see it as us."

When someone else is responsible for acting and the consequences, it gives us permission to ignore problems, even when we could be part of the solution. Science fiction author Douglas Adams, in his book *Life, the Universe and Everything*, offered a clever explanation of the power of making a difficulty somebody else's problem. His characters explained that rather than creating technology to make something invisible—which is difficult and extremely expensive—creating technology that surrounds it with a "Somebody Else's Problem Field" is much simpler and more effective. Once a difficult challenge becomes somebody else's problem, it is functionally invisible. I've always remembered this clever bit of writing because I've seen it happen many times in public policy. We are too quick to outsource environmental action to the government.

When people feel an issue has been handled, research shows they feel little need to take personal action. Over the course of a year, researchers at the University of Michigan studied the pro-environmental behavior of 600 people who had varying beliefs about climate change.[316] They found that those who "were most supportive of government climate policies," were also the "least likely to report individual-level actions." By way of contrast, those who "opposed government solutions," were "most likely to report engaging in individual-level pro-environmental behaviors." For some, government spending became a surrogate for personal environmental action. It doesn't mean those who rely on government don't care. Nor is it an argument against government action. Instead, it is an important reminder that handing a problem to an agency or politicians can provide a false sense of security, which offers individuals the license to focus on other issues on the assumption that an issue has already been addressed.

How we think about environmental action is influenced by objective and subjective information. First, there are tangible results. Is CO_2 declining? Are endangered species recovering? These are measurable (with some margin of error) and can provide the basis for accountable and adaptive actions. Second, there is a more personal sense of purpose—the feeling you are part of an important effort to help the planet. While motivating, when purpose becomes righteousness it can become counterproductive. Such feelings are hard to quantify and can be bought cheaply with strident language and dramatic policy proposals. At its best, a sense of purpose motivates action that creates tangible results. At its worst, it substitutes words for results.

In the social media age, this dynamic can become even worse. Without responsibility for the results of a public policy, people feel free to choose positions that, while of dubious merit, maximize benefits to their image. Similarly, the less leaders are judged by real-world results, the more reckless policy recommendations become and the greater the influence of public perception. The most remarkable experience I had with the power of public image to overwhelm environmental benefit occurred several years back. After the Spokane, Washington, school district built several new schools to "green" building standards, I gathered utility data from the school district and compared the energy use of the green schools to that of schools built recently without the green elements. I found that the performance of the so-called "green" schools was only marginally better, and the most energy-efficient school was the ironically named nongreen Browne Elementary. This counterintuitive result caught the eye of the education reporter for the local newspaper, who interviewed me about the findings. Less than an hour later the reporter called me back and said nervously, "I have a problem." The school district, he explained, said they had not heard of me and told the reporter that I had never requested any data from them. Further, they continued, the data I provided appeared to be completely made up. I told the reporter that I would fax over the data provided by the school district. For those who aren't familiar with fax machines (or whose memories have faded), it was customary to send a cover sheet first, providing information about who sent the fax and how many pages would follow. With my fax, I included the cover sheet sent by the district, which was on Spokane School District letterhead. A few moments after I sent the fax, my phone rang again. It was the reporter. His tone had changed. He told me that the person who signed the school district's cover sheet was the same person who told the reporter that I had never requested the data. When the reporter subsequently showed the fax cover sheet to the district official, they admitted that they were wrong.

Rather than admit that the green schools weren't performing and committing to find ways to improve their current and future performance—as would be appropriate if the goal was to reduce environmental harm—the district initially denied anything was wrong and attacked the messenger. Avoiding embarrassment was more important than fixing the problem. The district's incentives were aligned with preserving their public image rather than helping the environment.

Perhaps most interesting is what happened next. The story appeared but without any mention of the district's initial denials. This was galling because had I not saved the fax cover sheet, they might have gotten away with a false accusation. Additionally, it meant there was no accountability for the district's recklessness. Given a choice between maintaining a relationship with me or school district officials, the local reporter chose district officials he had to work with in the future. His incentives were about his work, not telling the full story about an environmental failure and the effort to hide it.

While I was annoyed at the behavior of district officials, the truth is that we can't expect people to consistently act contrary to their own interests and reward systems. If we punish people for experimentation and the resulting failures that are a byproduct of trying new things, then people will stop experimenting and will hide failures. Likewise, if people are not held accountable for failure to achieve environmental goals but are rewarded for efforts that sound good, there will be a lot of cool-sounding environmental programs but few environmental successes.

Recognizing these flaws, some school districts and builders have changed how green buildings are managed. Rather than building to meet a "green" construction checklist and walking away, districts and private companies now have performance contracts, paying builders and building managers a share of energy savings rather than promises that may never pan out. There is a balance (turning off the heat in the winter saves energy but makes for frozen school-children), but it makes it more likely that environmental promises will turn into real improvements. The types of technology making that possible for school districts and large companies are now available for individual homes. Small, personal technologies shift the balance of forces away from subjective assessments of intent toward measurable outcomes. The more power we put in the hands of individuals—whether they are at home or at work—the less we have to hope politicians have incentives to make the right choices.

TAKING ENVIRONMENTALISM SERIOUSLY

Connecting people to outcomes also makes it more likely that actions will reflect the seriousness of the environmental challenges we face. When something matters, people take steps to make sure they get what they want. I may want to lose weight, but when I eat those three pieces of pizza, I am making it clear that enjoying food matters more than my beach body. No matter how much I protest, the third piece of pizza holds the truth.

The question of motives becomes relevant when there aren't clear rewards for behaving in an environmentally positive way. Public policy debates too frequently devolve into questions about the motives of those who have different viewpoints because political incentives are complex and hard to discern. Is someone motivated by money, ideology, power, or their self-image as a good person, putting personal benefit ahead of stewardship of the planet? Only that person can know, and we can even fool ourselves about what truly drives us. Attacking the motives of political opponents is effective because such arguments are hard to rebut, and it is emotionally satisfying to tear down others to build ourselves up. Rather than address the merit of policies being offered, each side tries to destroy the credibility of political opponents. After all, why should we even listen to a "denier," an "alarmist," someone funded by "Big Oil" or George Soros?

Connecting people directly to environmental outcomes helps solve this problem. Providing a transparent and meaningful personal incentive to be a good environmental steward reduces the influence of competing interests like self-image. Someone might scoff at what they feel is a crisis mentality surrounding the risk of drought, but if technology helps cut their water bill, they will use the technology. It is hard to accuse someone of ulterior motives when they openly admit that caring for the planet is also personally rewarding. Incentive-based systems are more reliable than hoping for pure motives. When someone says, "You're only doing this because it saves you money," people can confidently and proudly answer, "Of course! Isn't it great?" The environment doesn't care why you are saving energy or protecting wildlife habitat. Salmon are just happy you are saving water, leaving more in streams for them to enjoy.

Small, environmental technologies aren't going to eliminate divisive, political rhetoric. But they can reduce the amount of environmental progress that is contingent on that type of rhetoric. Personal environmental technologies provide an alternative path for stewardship that doesn't involve slugging it out on Twitter. Many of the most innovative environmental approaches I've outlined are in places where political power is weak or counterproductive. In those circumstances, there isn't the temptation to focus on influencing government because it is unlikely to yield positive outcomes. In those circumstances, groups like Paso Pacifico in Nicaragua look for alternatives. That is what it means to take environmentalism seriously. When the primary metric of success is reducing harm to the environment rather than personal image, the incentives to engage in destructive political battles can be replaced with rewards for environmental stewardship.

A DYNAMIC AND EVOLUTIONARY APPROACH

To make the best use of the small environmental approach, we will need to change our mindset about the process of addressing threats to the planet. The idealized version of current environmental policy builds a public policy plan based on scientific knowledge. I say that is the "idealized" version because while many politicians claim to follow that path, most policies are influenced heavily by ideology, political realities, and other nonscience factors. While the planning approach works in some instances, we should not assume it is the best approach. Science-based planning can be effective when there is low uncertainty and the number of confounding influences is limited. For many problems, a more dynamic approach is necessary to deal with the reality of limited knowledge and complex ecological, economic, or social interactions.

The dynamic approach of solving environmental problems—one that utilizes a process of observation and adjustment—is not new. It resembles the approach used by North American tribal communities to manage natural resources, presettlement. The tribal use of fire in the western United States is a good example of this process.

For millennia, tribes used fire for hunting, habitat manipulation, traditional ceremonies, and forest management. Using fire required skill and became a well-developed practice. One history of the use of fire by tribes explains that "Salish-Kalispel elders have described how the application of fire was a difficult, complicated, and dangerous task, one only learned through long experience and entrusted to a person referred to as the *sxʷpaám* (pronounced spa · ám), the one who makes fire, a person of high knowledge and training."[317] That position of respect reflected the value many tribes put on the use of fire as a tool and the level of training required. The skill of the *sxʷpaám* was not the result of a series of rigorous scientific experiments. Tribal members learned from experience, adjusting to changing conditions and applying new techniques. Although the process was not scientifically robust, the knowledge they gained was better than existing science.

As settlers moved into tribal lands, they did not appreciate or understand the use of fire and the knowledge underlying these practices. As the state of Montana's 2020 Forest Action Plan noted, "non-Indigenous people generally assumed that tribal fire practices, and the cultures of which they were a part, were 'primitive' and at odds with 'progress.'" The US government began to criminalize tribal burning—sometimes using lethal force—and as a result, Indians lit fewer fires after the 1880s. After massive fires swept across the American West in 1910, the policy of the US government was to put out all forest fires as quickly as possible. The role of fire in maintaining healthy forests on both tribal and nontribal land was not replaced with active management to mimic fire disturbance. Millions of acres of dense forests of species that were vulnerable to insects, disease, drought, and fire resulted. Under tribal management practices, open, patchy forests of fire-resistant species were maintained by tribes for millennia. Fires—both natural and human-caused—created a mosaic of many different patches each with its own mix of tree species, ages, and sizes. With fire suppression, that patchwork became more homogeneous, more like a blanket than a quilt. This loss of structural diversity caused a corresponding loss of habitat diversity for wildlife and increased the risk of catastrophic loss from drought, insects, disease, and wildfire. Science is now catching up with the indigenous knowledge, and there is a strong consensus that controlled burns must be a part of strategies to return forests in the western United States to a healthier condition.[318]

The key to traditional knowledge is more than just the information itself. It is also the learning process. Although different from the scientific process, the knowledge and practices it generates are useful and important. That sentiment is echoed by the United Nations Educational, Scientific, and Cultural Organization (UNESCO). Their statement on "Sustainable Development and Environmental Change," says, "While science contributes significantly to understanding earth systems, social systems and their interactions, there is growing awareness that scientific knowledge alone is inadequate for solving the emerging environmental crises. The knowledge of indigenous peoples and local communities—often referred to as local, indigenous or traditional knowledge—is now recognized as essential, alongside science, for developing effective and meaningful action world-wide."[319] A tribal scientist I work with told me that while scientific rigor is important, it often seeks "universal context-free truths." When facing real-world problems, "we are not working in labs. There are an awful lot of things that you can't anticipate or know about. When you look at traditional knowledge you understand that things are not context free."

The approach of dynamically generating and applying knowledge is similar to the philosophy of Nobel Prize–winning economist Friedrich von Hayek. He argued that the knowledge necessary to completely understand complex systems is unobtainable. Applying Hayek's ideas to fighting climate change, retired philosophy professor Mark Sagoff argued this approach is "a discovery problem, that is, a problem of how people may act not on the knowledge they have but on the knowledge they find, create, and apply to change both their wants and the resources they need to satisfy those wants."[320] This evolutionary process of trial and error is what makes the small-technology approach to environmental action so powerful. In the same way that the process of traditional knowledge complements scientific knowledge, individual environmental actions supplement government policy. Politicians frequently assure the public that they are "following the science" when making environmental policy. In many cases, that claim is little more than rhetoric. It also betrays a static understanding of both knowledge and the environmental challenges we face. Rather than using "the" science to create a long-term plan, the approach suggested by traditional knowledge and small technology is to experiment, learn what works, and adjust to the consequences of our actions, intended and unintended. The low cost of innovation and information accelerates the trial-and-error process, providing feedback that is more robust and strengthening the ability to adjust. Reducing the barriers to innovation also reduces the cost of failure. People can experiment, learning from failures and applying the knowledge to the next iteration of experimentation.

It took decades for the US Forest Service to appreciate the traditional knowledge of prescribed burning and decades more for it to become a meaningful part of government policy, in part because the evidence for controlled burning from tribal experience was perceived to be based in tradition rather than science. There are traces of this same conceit in how we make environmental policy today. The lived experience of those who have knowledge and a stake in sustainable outcomes can be extremely powerful and a useful guide to good practice.

HOW POLICY MAKERS CAN DEMOCRATIZE ENVIRONMENTALISM

Many of the small technologies outlined here have emerged organically, independent of public policy, in a dynamic sometimes called "permissionless innovation." Other technologies are reactive to public policies that price electricity, promote citizen science, or encourage the use of innovative approaches. To help accelerate the growth of small, environmental technologies, elected officials and agency staff should do three things.

The most basic step policy makers can take is to remove barriers to innovation and actively include inventive environmental approaches. Regulatory barriers in the West, often a holdover from previous decades when a top-down approach was the best option, can make it difficult to take advantage of new technology.

If we reduce the regulatory barriers, we will also have to shift the focus of accountability when things go wrong from government agencies to innovators. When consumers have a bad experience or there is pollution in the environment, some immediately ask, "Why didn't government prevent this?" Over time, this results in regulations that reduce risk but also reduce innovation. In extreme versions it can shut down innovation completely. There is actually a name for this extremely cautious approach—"nestoring"—after John Nestor, who worked at the Food and Drug Administration and was in charge of approving renal and cardiac drugs. Between 1968 and 1972, he failed to approve a single new treatment. Medicines that aren't approved can't harm anyone, he reasoned.[321] They also can't help people. A cautious approach that prevents potential harm also inhibits projects that offer environmental benefits. We can hope that agency staff will resist the pressures to choose a path of safe inactivity and take risks to support efforts that are good for the economy and environment. It is true that many public servants are motivated by a desire to make their community better and do their job in the best way they can. However, a system that relies on the good motives of its employees to take risks in favor of action while providing incentives to be cautious is inherently fragile. Humans are loss-averse—we feel the pain of loss more than the joy of gain. This is even more true when agency staff don't get to enjoy the benefits of projects that move forward but do feel the pain of being held accountable for projects that fail. People are willing to sacrifice for a larger principle, but the more we ask people to sacrifice themselves, the more likely it is that public servants will wonder whether it is worth it and begin to nestor.

To encourage more dynamism, we will have to acknowledge these risks, provide more leeway to innovators but also hold them responsible for failure. If staff at government agencies are going to be held accountable for environmental harm, they will, quite rationally, work to prevent actions that might reflect badly on them. Some see those restrictions as justifiable because they reduce environmental harm. What often gets lost is the opportunity cost of those regulations—the environmental projects and protection that doesn't occur because of excessive regulation. It is a hard balance to find, to be sure. The amazing innovations that have already occurred demonstrate the possibilities. More such opportunities are out there if we plant the seeds of innovation and allow them to flourish.

Second, policy makers should focus on goals, not methods. Many environmental policies and expenditures focus on particular ways of achieving a goal. Schools must build using a green building checklist, rather than measuring actual energy use. Legislatures subsidize particular biofuels rather than adopting policies that reward CO_2 reductions no matter how they occur. By focusing on goals, we encourage experimentation to discover the most effective approaches to help the planet. Just as important, it allows bad ideas to fail and be replaced with superior innovations. When existing regulations favor particular technologies, competing innovations must not only perform better; they must also appeal to politicians or bureaucrats to overcome political hurdles. Ecosystems only care about outcomes. We should too.

Finally, we should encourage a culture of transparency. The simple act of making price data available on an API caused people to create tools that helped Octopus Energy customers save money and energy. From Rainforest Connection's collection of sound data to create a weather station for biodiversity to Ecobee's voluntary data-sharing program, making information available engages the creativity of activists and data scientists who want to put it to work for the environment. I was able to determine that green schools didn't always perform well because the utility data for the school buildings was publicly available. Improving access to data makes it easier to identify what works and expose what doesn't.

Publicly available data also helps ground-truth political debates. Transparent and reliable data can focus discussion on areas of real disagreement rather than shouting matches where all arguments are labeled "fake news" by the other side. I have saved dozens of links to environmental data in my web browser. When people debate about the extent of sea ice in the Arctic or Antarctic, I call tell you the status virtually instantly.[322] What percentage of California's power came from renewable sources yesterday? I can tell you that.[323] People can argue about causation, but thermometers make it clear that global temperatures are increasing.[324, 325]

Environmental data helps engage the public, calibrating our response to environmental problems and creating the information to address those challenges. Policy makers should make more environmental data available. Some steps have already been taken in this direction. In the United States, the Crowdsourcing and Citizen Science Act requires that "a Federal science agency shall, where appropriate and to the extent practicable, make data collected through a crowdsourcing or citizen science project under this section available to the public, in a machine readable format, unless prohibited by law."[326] While ensuring privacy is protected, making government data available opens up new opportunities for environmental awareness and action.

These recommendations are targeted primarily to developed countries. There will be challenges in applying these principles and use will vary from place to place. When those challenges arise, there should be a bias toward democratizing environmental efforts and encouraging innovation.

A CONSERVATION REVOLUTION

One of my favorite sayings is that the man who says it can't be done should get out of the way of the woman who is doing it.

As I have shown in the preceding pages, there are many women (and men) who are pioneering a new approach to environmental sustainability. One of them is Talia Speaker, who leads the research program for **WILD**LABS, an NGO collaborative that supports the use and development of conservation technology. She told me she was interested in conservation in college, but, she said, "All of my conservation fieldwork was terrible and really slow. Our tools were really clunky. And I thought my iPhone can do so much, why do we not have better tools for such important work?" **WILD**LABS' mission is to help enable the creation and use of those tools.

In August 2021, in partnership with environmental technology platforms EarthRanger and Skylight, **WILD**LABS surveyed the global conservation community to understand the role technology plays in their work to protect wildlife. They found that "even before the pandemic, technological innovations were rapidly enhancing our ability to preserve our planet, ecosystems, and local economies."[327] As the COVID pandemic contributed to an increase in poaching and other illegal activity, conservation technology became more important as conservation NGOs struggled with fewer resources. "Pressure and challenges can spur innovation, which we have certainly witnessed both globally and within the conservation community over the last eighteen months," the survey authors noted.

Despite the challenges, the survey found that more than half of respondents reported feeling "more optimistic" about the role conservation technology can play in their work, with only 7 percent feeling less optimistic. "The greatest drivers of this optimism are the increasing accessibility of conservation technologies and the rate at which the field is evolving," the survey authors reported. "This positive outlook toward conservation technology is also evident in the continued adoption of new technology solutions across wide-ranging needs—from tools to aid data collection in the field, to resources that will help them analyze that data effectively. With limited bandwidth for personnel, technology plays an important role in tackling different challenges, accomplishing tasks and filling gaps in capacity."

Ultimately, two-thirds of survey respondents indicated they would use conservation technology more in the future.

Those results are encouraging to Talia Speaker. The emergence of a diverse set of technology tools "is giving conservationists a fighting chance, though these tools still have a really long way to go in realizing that potential," she told me. The technologies alone aren't enough. **WILD**LABS is working to create a support network, including training conservationists on using technology, helping maintain the equipment, and connect with experts who can create new tools. "You don't just need the tool, you need a support system to adopt this tool," she said. "It is more complicated than just empowering a bunch of little efforts." Speaker, **WILD**LABS, and the many other innovators I've mentioned are creating that ecosystem of support. The impact is much more than practical—it is building an ethic of empowerment that is an important part of realizing the power of small, environmental technologies.

Darlene Cavalier, the founder of SciStarter we met in chapter 8, highlighted the connection between a democratized ethic of responsibility for making the world a better place and innovation. Upon her appointment as cochair for innovation, science, and entrepreneurship for the US semiquincentennial (250-year) celebration, she and her cochair noted that "the concurrence in the late eighteenth century of scientific and democratic revolutions was no coincidence—they were integrally linked phenomena put into action by many of the same people. At their best, entrepreneurship and innovation follow in the same Enlightenment tradition, refusing to take the world as we find it as the best the world can be."[328] It is no accident that making environmental innovation accessible to individuals is coinciding with an explosion of technologies that promote sustainability and conservation. That spirit of making the world better by unleashing bottom-up innovation is spawning an inclusive approach to environmental action, one that welcomes the contributions of technology-empowered individuals. It holds the promise of revolutionizing the way we care for the planet.

ACKNOWLEDGMENTS

I have been developing this book for several years, and there are many along the way who were extremely helpful—cheering me on, helping me think through these issues, and spending time to talk about their work without promise of anything other than the hope they were spreading good ideas and making the world better.

There are many rough patches when writing a book. Several people helped me through those moments. My wife propped me up repeatedly. My sister-in-law Anna edited several early chapters, offering her English skills and encouragement. My sister and parents patiently listened to my griping. It all helped, and I am very appreciative.

I am also thankful for my colleagues at the Washington Policy Center. As these ideas developed over several years, they talked to me about them and worked through various iterations of the issues. They provided the intellectual freedom for me to explore ideas that sometimes seemed half-baked. Mariya Frost was particularly encouraging and offered on more than one occasion to harm anyone who was causing me problems. Fortunately, I didn't have to resort to that.

Thanks also to my colleagues around the country who offered honest feedback and support. Thanks to Jennifer Butler, who asked questions and was interested in these arguments before many others. I appreciate my colleagues at other State Policy Network organizations, including Jason Hayes, Isaac Orr, Dave Stevenson, Amy Oliver Cooke, and others. The people at the Property and Environment Research Center, including Holly Fretwell, Hannah Downey, Shawn Regan, Jonathan Wood, and Brian Yablonski, are a constant inspiration—committed to finding innovative approaches to environmental stewardship. And to Yoram Bauman, who called the topic of my book "wacky" when I first told him about it but was always supportive and ended up saying nice things.

The final form of *Time to Think Small* is very different from the first draft, in large part due to the thoughts, reactions, and suggestions of my agent, Regina Ryan, and editor Alexandra Halsey. They offered different perspectives and helped me write for people independent of political perspective. And special thanks to Regina, who kept fighting to get the book published long after others would have given up on it.

And then the people who influence you without realizing it. Russ Roberts, whose podcast introduced some of the concepts I've included. David Troutt, who challenged me even when he wasn't intending to. Pat Coussens, who understood my vision, encouraged it, and was always available to edit my drafts.

Thanks to the staff at Aroma Coffee in Fall City, Washington. They let me live the writer's life of drinking coffee and working on the book for hours.

Finally, I appreciate everyone who reads and considers my argument. Of course, I look forward to hearing from those who enjoy the book and are inspired by it. But I also welcome sincere critiques that will lead me to update and improve my thinking in the upcoming years.

ENDNOTES

CHAPTER 1

1. William Ruckelshaus, "A New Shade of Green," *Wall Street Journal*, April 17, 2010, https://www.wsj.com/articles/SB1000142405270230341040457515164 0963114892.

2. Bill and Melinda Gates, "Our 2019 Annual Letter: We Didn't See This Coming," February 12, 2019, https://www.gatesnotes.com/2019-Annual-Letter.

3. Todd Myers, "Solar Subsidies Take Money from the Poor to Help the Rich," *Wall Street Journal*, November 14, 2013, https://blogs.wsj.com/experts/2013/11/14/solar-subsidies-take-money-from-the-poor-to-help-the-rich/.

4. Lazard, "Levelized Cost of Energy and Levelized Cost of Storage 2019," November 7, 2019, https://www.lazard.com/perspective/lcoe2019/.

5. Todd Myers and Steve Sexton, "A Contrarian View of the Local Food Movement," *Wall Street Journal*, July 16, 2015, https://blogs.wsj.com/experts/2015/07/16/a-contrarian-view-of-the-local-food-movement/.

6. Ministry of Environment and Food of Denmark, "Life Cycle Assessment of Grocery Carrier Bags," Environmental Project No. 1985, February 2018, https://www2.mst.dk/Udgiv/publications/2018/02/978-87-93614-73-4.pdf.

7. UK Environment Agency, "Life Cycle Assessment of Supermarket Carrier Bags: A Review of the Bags Available in 2006," February 2011, https://assets.publishing.service.gov.uk/government/uploads/system/uploads/attachment_data/file/291023/scho0711buan-e-e.pdf.

8. Elinor Ostrom, *Governing the Commons: The Evolution of Institutions for Collective Action*, Cambridge University Press, 1991.

9. Brad Smith, *Tools and Weapons: The Promise and Peril of the Digital Age*, New York: Penguin, 2019, p. 289.

10. Elaine Woo, "Elinor Ostrom Dies at 78; First Woman to Win Nobel in Economics," *Los Angeles Times*, June 13, 2012, http://articles.latimes.com/2012/jun/13/local/la-me-elinor-ostrom-20120613.

11. Todd Myers, "The Environmental Failure of 'Green' Schools," *Wall Street Journal*, November 17, 2015, https://blogs.wsj.com/experts/2015/11/17/the-environmental-failure-of-green-schools/.

12. Bret Stephens, "Climate of Unintended Consequences," *New York Times,* May 4, 2017, https://www.nytimes.com/2017/05/04/opinion/climate-policy-ethanol.html.

13. EcoWatch, "5 Companies Leading the Charge in Using Ocean Plastic in Their Products," June 4, 2016, https://www.ecowatch.com/5-companies-leading-the-charge-in-using-ocean-plastic-in-their-product-1891161554.html.

14. Lisa Stiffler and Chris McGann, "Bill Orders Firm Steps to Make State 'Greener,'" *Seattle P-I,* February 19, 2008, https://www.seattlepi.com/local/article/Bill-orders-firm-steps-to-make-state-greener-1264919.php.

15. Frank Newport, "New Series: Where Americans Stand on the Environment, Energy," March 22, 2018, https://news.gallup.com/opinion/gallup/231386/new-series-americans-stand-environment-energy.aspx.

16. Tyler Cowen, *The Great Stagnation,* New York: Penguin, 2005, loc. 575 of 988, Kindle.

17. Benjamin W. Nelson et al.,"Wild Chinook Salmon Productivity Is Negatively Related to Seal Density and Not Related to Hatchery Releases in the Pacific Northwest," *Canadian Journal of Fisheries and Aquatic Sciences* 76 (2019): 447–462, https://doi.org/10.1139/cjfas-2017-0481.

CHAPTER 2

18. National Honey Board, "Honey Trivia," https://www.honey.com/newsroom/presskit/honey-trivia (Accessed May 9, 2020).

19. United Nations, "Water and Sanitation—United Nations Sustainable Development," https://www.un.org/sustainabledevelopment/water-and-sanitation/ (Accessed May 9, 2020).

20. Amber Donovan-Stevens, "Tackling the African Water Crisis with Mobile Technology," *Chief Sustainability Officer,* June 3, 2019, https://www.csomagazine.com/sustainability/tackling-african-water-crisis-mobile-technology.

21. World Health Organization, "Drinking-water," June 14, 2019, https://www.who.int/en/news-room/fact-sheets/detail/drinking-water.

22. Everett M. Rogers, *Diffusion of Innovations,* New York: Free Press, 2003, p. 108.

23. Anjalee Burr, "Hand Pumps Are Dinosaurs—Why Use IBM Technology on Pre-historic Hardware," eWaterPay, January 27, 2020, http://ewaterpay.com/hand-pumps-are-dinosaurs-why-use-ibm-technology-on-pre-historic-hardware/.

24. Amazon, "eWater Case Study," 2017, https://aws.amazon.com/solutions/case-studies/ewater/ (Accessed May 10, 2020).

25. Donovan-Stevens, 2019.

26. eWaterPay, "'We Give Water USA' partner with eWaterPay to rehabilitate Wellingaraba, The Gambia," January 16, 2020, https://ewaterpay.com/we-give-water-partner-with-ewaterpay-to-rehabilitate-wellingaraba-the-gambia/.

27. WaterAid, "One Token Changing the Game for Sustainable Rural Water Supply in Tanzania," January 31, 2018, https://washmatters.wateraid.org/blog/one-token-changing-the-game-for-sustainable-rural-water-supply-in-tanzania.

28. eWaterPay, "Global—Commercial Dashboards," https://commercial.ewater.services/ (Accessed May 21, 2021).

29. Rogers, 2003, p. 49.

30. Alison George, "Why Believe in a Flat Earth?" *New Scientist* 229, no. 3059 (February 6, 2016): 25, https://doi.org/10.1016/S0262-4079(16)30274-3.

31. US Centers for Disease Control and Prevention, "Global Diarrhea Burden," https://www.cdc.gov/healthywater/global/diarrhea-burden.html (Accessed May 10, 2020).

32. Caroline Haskins, "How Ring Went from 'Shark Tank' Reject to America's Scariest Surveillance Company," *Vice*, December 3, 2019, https://www.vice.com/en_us/article/zmjp53/how-ring-went-from-shark-tank-reject-to-americas-scariest-surveillance-company.

33. Rogers, 2003, p. 221.

34. Bill Henderson, "Fast Versus Slow Innovations (011)," June 21, 2017, https://www.legalevolution.org/2017/06/fast-versus-slow-innovations-011/.

35. Caitlin Drummond and Baruch Fischhoff, "Science Knowledge and Polarization," *Proceedings of the National Academy of Sciences*, August 2017, 201704882, https://doi.org/10.1073/pnas.1704882114.

36. For example, compare the difference between William Nordhaus, who won the Nobel Prize in economics for his climate modeling (https://www.nobelprize.org/prizes/economic-sciences/2018/nordhaus/lecture/), and former chief economist for the World Bank Nicholas Stern (http://www.lse.ac.uk/GranthamInstitute/publication/the-economics-of-climate-change-the-stern-review/).

37. Rogers, 2003, p. 332.

38. Indiegogo, "The Seabin Project," https://www.indiegogo.com/projects/cleaning-the-oceans-one-marina-at-a-time#/ (Accessed October 6, 2019).

39. Salmon Innovation Fund, "Home | Salmonfund," https://www.salmonfund.com/.

40. Seabin, "Safe Harbor Marinas North America," https://seabinproject.com/marina/safe-harbor-marinas/ (Accessed October 6, 2019).

41. Seabin Project, "Seabin Project—Cleaner Oceans for a Brighter Future," https://seabinproject.com/ (Accessed May 21, 2021).

42. Andrea James, "Tully's Discards Leaky Compostable Cups, for Now," *Seattle*

P-I, May 22, 2009, https://blog.seattlepi.com/thebigblog/2009/05/22/tullys-discards-leaky-compostable-cups-for-now/.

43. Rogers, 2003, p. 283.

44. Heather Corcoran and Danny White, "New Features on Kickstarter Encourage Creators to Think Green," November 27, 2018, https://medium.com/the-fourth-wave/new-features-on-kickstarter-encourage-creators-to-think-green-b17a05f41bab (Accessed October 7, 2019).

45. Kickstarter.com, "Shapeshift: An Open Call for Projects," https://creators.kickstarter.com/shapeshift/?ref=section-homepage-promo-shapeshift (Accessed October 7, 2019).

46. Environmental Defense Fund, "Natural Gas | EDF+Business," http://business.edf.org/projects/featured/natural-gas (Accessed October 7, 2019).

47. Marc Gunther, "Crowdsourcing Green," October 2, 2011, https://web.archive.org/web/20170301222728/http://www.marcgunther.com/crowdsourcing-green/.

48. JikoPower, "JikoPower Spark: Charge Your Cell Phone with Fire by Jiko-Power," Kickstarter.com, https://www.kickstarter.com/projects/1072708080/jikopower-spark-charge-your-cell-phone-with-fire/description (Accessed October 18, 2019).

49. Seabin.com, "Seabin Project Tackling Microfibers Head On," https://seabinproject.com/seabin-project-tackling-microfibers-head-on/ (Accessed October 20, 2019).

50. Karen Weintraub, "Elephants Are Very Scared of Bees. That Could Save Their Lives," *New York Times*, January 26, 2018, https://www.nytimes.com/2018/01/26/science/bees-elephants-.html.

CHAPTER 3

51. Sense.com, "How a Texas Resident Used Sense Data During the Rolling Blackouts," https://blog.sense.com/how-a-texas-resident-used-sense-data-during-the-rolling-blackouts/.

52. Abby Ellin, "How Skipping Hotel Housekeeping Can Help the Environment and Your Wallet," *New York Times*, February 27, 2018, https://www.nytimes.com/2018/02/27/travel/skipping-hotel-housekeeping-perks.html (Accessed December 29, 2018).

53. Cass R. Sunstein and Richard H. Thaler, *Nudge: Improving Decisions about Health, Wealth, and Happiness*, New York: Penguin Press, 2008.

54. Hunt Alcott and Todd Rogers, "The Short-Run and Long-Run Effects of

Behavioral Interventions: Experimental Evidence from Energy Conservation," *American Economic Review* 104, no. 10 (October 2014): 3003–37, https://www.jstor.org/stable/43495312.

55. David Nemtzow (director of the Building Technologies Office, US Department of Energy), phone interview with author, November 5, 2018.

56. Ito Koichiro, Takanori Ida, and Makoto Tanaka, "Moral Suasion and Economic Incentives: Field Experimental Evidence from Energy Demand," *American Economic Journal: Economic Policy*, no. 1 (2018): 240–67, https://doi.org/10.1257/pol.20160093.

57. Nassim Taleb and Constantine Sandis, "The Skin in the Game Heuristic for Protection against Tail Events," July 30, 2013, http://www.primequadrant.com/wp-content/themes/PrimeQuadrant_1.0/library/docs/Essay_TheSkinInTheGame.pdf.

58. Barack Obama, "Remarks by the President on Recovery Act Funding for Smart Grid Technology," October 26, 2009, https://obamawhitehouse.archives.gov/photos-and-video/video/president-obama-explains-smart-grid-and-economic-recovery#transcript.

59. Naperville Smart Meter Awareness v. City of Naperville, No. 11 C 9299 (7th Cir. 2018).

60. Philip E. Tetlock, *Superforecasting*, New York: Crown Publishers, 2015, p. 14.

61. Steve Barager, phone interview with author, February 12, 2016.

62. Watersmart, "Connecting the Dots for Consumers," June 6, 2017, https://thirsty.watersmart.com/blog/connecting-dots (Accessed March 2, 2019).

63. Marc de Jong and Menno van Dijk, "Disrupting Beliefs: A New Approach to Business-Model Innovation," *McKinsey Quarterly*, July 2015, https://www.mckinsey.com/business-functions/strategy-and-corporate-finance/our-insights/disrupting-beliefs-a-new-approach-to-business-model-innovation.

64. Chris King and Bonnie Datta, "Action by Choice: Time-Varying Rates Is an Effective Way to Satisfy Customer Demands," *Public Utilities Fortnightly*, December 2015, https://www.fortnightly.com/fortnightly/2015/12/action-choice.

65. Edward Cazalet, phone interview with author, February 12, 2016.

66. Herman K. Trabish, "The New Demand Response and the Future of the Power Sector," *Utility Dive*, December 11, 2017, https://www.utilitydive.com/news/the-new-demand-response-and-the-future-of-the-power-sector/512134/.

67. Adam Uzialko, "Smart Energy: Using IoT and AI to Reduce Waste, Boost Profits," *Business News Daily*, July 10, 2017, https://www.businessnewsdaily.com/10064-energy-management-iot-ai-machine-learning.html.

68. US Environmental Protection Agency, letter to "Programmable Thermostat Manufacturer or Other Interested Stakeholder," May 4, 2009, https://www.energystar.gov/ia/partners/prod_development/revisions/downloads/thermostats/Spec_Suspension_Memo_May2009.pdf?ddc6-9ea1.

69. Oscar Gans, Florida Power & Light Co., letter to Katherine Kaplan, acting branch chief, ENERGY STAR Product Labeling, https://www.energystar.gov/ia/partners/prod_development/revisions/downloads/thermostats/Florida-Power-Light-Comments.pdf?a271-cf42 (Accessed March 3, 2019).

70. Tom Simonite, "How Nest's Control Freaks Reinvented the Thermostat," *MIT Technology Review*, February 15, 2013, https://www.technologyreview.com/s/511086/how-nests-control-freaks-reinvented-the-thermostat/.

71. Robert Walton, "5 Trends to Watch in Utility Customer Management," *Utility Dive*, June 19, 2017, https://www.utilitydive.com/news/5-trends-to-watch-in-utility-customer-engagement/444627/.

72. Marissa Hummon, *Tendril*, November 2, 2016.

73. US Energy Information Administration, "Electricity Data Browser," https://www.eia.gov/electricity/data/browser/ (Accessed May 21, 2021).

74. Portland General Electric, letter to Public Utility Commission of Oregon, "PGE's Application for Deferral of Expenses Associated with Two Residential Demand Response Pilots," October 3, 2014, https://edocs.puc.state.or.us/efdocs/HAA/um1708haa134559.pdf.

75. Cadmus, memorandum to Josh Keeling and Alex Reedin, Portland General Electric, "PGE Rush Hour Rewards Findings Summary," December 27, 2016.

76. Public Utility Commission of Oregon, "Order No. 17-244," July 11, 2017, https://apps.puc.state.or.us/orders/2017ords/17-244.pdf.

77. Nest, "It's Time for Time of Savings," June 21, 2016, https://nest.com/blog/2016/06/21/its-time-for-time-of-savings/.

78. Nexant and Research into Action, "California Statewide Opt-in Time-of-Use Pricing Pilot," March 30, 2018, http://www.cpuc.ca.gov/WorkArea/DownloadAsset.aspx?id=6442457172.

79. Marc Reisner, *Cadillac Desert*, https://www.amazon.com/dp/B001RTKIUA/ref=dp-kindle-redirect?_encoding=UTF8&btkr=1 (Accessed March 9, 2019).

80. Adele Peters, "Water Bills Are Going Up, This Device Helps You Lower Them," *Fast Company*, November 21, 2017, https://www.fastcompany.com/40498687/water-bills-are-going-up-this-device-helps-you-lower-them.

81. Phyn, "Technology," https://www.phyn.com/technology/ (Accessed March 30, 2019).

82. Flume, "Flume Water | Smart Home Water Monitor | Water Leak Detector,"

https://flumewater.com/.

83. Resideo, "Buoy Discontinuation FAQs (resideo.com)," July 23, 2021, https://www.resideo.com/us/en/support/buoy-discontinuation-faqs/.

84. Keri Waters, video interview with author, August 25, 2021.

85. Irrigation Scheduler Mobile, "About Us," http://weather.wsu.edu/ism/index.php?m=1&action=about-us (Accessed March 24, 2019).

CHAPTER 4

86. US Energy Information Administration, "Levelized Cost and Levelized Avoided Cost of New Generation Resources in the Annual Energy Outlook 2019," February 2019, https://www.eia.gov/outlooks/aeo/pdf/electricity_generation.pdf.

87. Cliff Mass and Luke Madaus, "Surface Pressure Observations from Smartphones: A Potential Revolution for High-Resolution Weather Prediction?" *Journal of the American Meteorological Society* (September 2014): 1343–49, https://journals.ametsoc.org/doi/pdf/10.1175/BAMS-D-13-00188.1.

88. Callie McNicholas and Clifford F. Mass, "Bias Correction, Anonymization, and Analysis of Smartphone Pressure Observations Using Machine Learning and Multi-Resolution Kriging," *Weather and Forecasting* (published online ahead of print 2021), https://doi.org/10.1175/WAF-D-20-0222.1.

89. George Akerlof, "The Market for 'Lemons': Quality Uncertainty and the Market Mechanism," *Quarterly Journal of Economics* 84, no. 3 (August 1970): 488–500.

90. Galaxy Zoo, https://www.zooniverse.org/projects/zookeeper/galaxy-zoo/classify (Accessed May 29, 2019).

91. Nielsen, p. 135.

92. Weather Rescue, "Research," https://www.zooniverse.org/projects/edh/weather-rescue/about/research (Accessed May 29, 2019).

93. Earthquake Detective, "Research," https://www.zooniverse.org/projects/vivitang/earthquake-detective/about/research (Accessed May 29, 2019).

94. eBird.org, "Staff Bios," https://ebird.org/about/staff (Accessed June 20, 2018).

95. eBird.org, "Real-Time Checklist Submissions," https://ebird.org/livesubs (Accessed June 20, 2018).

96. eBird.org, "eBird Avicaching," https://ebird.org/science/applied-projects/avicaching (Accessed June 20, 2018).

97. Making Space for Earth, "Bidding for the Benefits of Birds," October 30, 2017, https://science.nasa.gov/earth-science/applied-sciences/making-space-for-earth/bidding-for-the-benefit-of-birds.

98. Eric Hallstein and Matthew L. Miller, "A Renter's Market: BirdReturns Offers

Innovative Conservation," Nature.org, August 6, 2014, https://blog.nature.org/science/2014/08/06/birds-birdreturns-innovative-lands-conservation-science/.

99. Catherine Wolfram, "Can Cell Phones Help Improve Electricity Reliability?" *Energy Institute Blog*, UC Berkeley, February 18, 2019, https://energyathaas.wordpress.com/2019/02/18/can-cell-phones-help-improve-electricity-reliability/.

100. Energy and Economic Growth, "GridWatch | EEG (energyeconomicgrowth.org)," https://energyeconomicgrowth.org/node/191 (Accessed August 24, 2021).

101. "DumsorWatch—PowerWatch Sensor," https://dumsorwatch.com/ (Accessed August 24, 2021).

102. Energy Information Administration, "How Much Energy Is Consumed in US Residential and Commercial Buildings?" May 14, 2019, https://www.eia.gov/tools/faqs/faq.php?id=86&t=1.

103. "Frequently Asked Questions for the Green New Deal" was released on February 7, 2019, and can be found at https://assets.documentcloud.org/documents/5729035/Green-New-Deal-FAQ.pdf (Accessed May 31, 2019).

104. Meta Brown, "Massive New Energy Use Data Resource Coming This January," *Forbes*, November 30, 2016, https://www.forbes.com/sites/metabrown/2016/11/30/massive-new-energy-use-data-resource-coming-this-january/#f0f8b6965df1.

105. Trang Be, "Is Your Home A Lemon?" Ecobee, July 30, 2018, https://www.ecobee.com/2018/07/is-your-home-a-lemon/.

106. Ecobee, "Donate your Data," https://www.ecobee.com/donateyourdata/.

107. Meredith Fowlie, Michael Greenstone, and Catherine Wolfram, "Do Energy Efficiency Investments Deliver? Evidence from the Weatherization Assistance Program," Energy Policy Institute at the University of Chicago, January 29, 2018, https://epic.uchicago.edu/sites/default/files/UCH-1605_CostofResidentialEEResearchSummary_v03_PressReady_0.pdf.

108. Fishackathon, "Fishackathon," https://fishackathon.co/ (Accessed June 12, 2019).

109. US Department of State Office of Global Partnerships, "Fishackathon 2016 Problem Statements," https://challenges.s3.amazonaws.com/fishackathon/2016%20Fishackathon%20Final%20Problem%20Statements.pdf (Accessed June 12, 2019).

110. HackerNest, "Fishackathon 2018 Challenge Statements," https://www.uma.es/media/files/Fishackathon_2018_Challenge_Statements_.pdf (Accessed June 12, 2019).

111. HackerNest and US Department of State, "Fishackathon 2018 Challenges," https://docs.google.com/document/d/1Rv7dmlvJyXxqkDgcvh8iG2k VPKzJ7Ee5TgG4ddDJi9U/edit (Accessed June 23, 2019).

112. Finnder, "Finnder—San Francisco Fishackathon 2018, 1st Place Winner," https://www.youtube.com/watch?v=0bp-LBeKwmI (Accessed June 23, 2019).

113. Hackernest Global, "Reasons To *Not* Run Hackathons (from Experts)," February 18, 2019, https://www.hackernest.com/hackathons.

CHAPTER 5

114. Environmental Protection Agency, "Air Pollutant Emissions Trends Data," https://www.epa.gov/air-emissions-inventories/air-pollutant-emissions-trends-data (Accessed June 5, 2020).

115. United States Bureau of Transportation Statistics, "US Vehicle-Miles," https://www.bts.gov/content/us-vehicle-miles (Accessed June 5, 2020).

116. Alain Bertaud, *Order without Design: How Markets Shape Cities,* Cambridge, Massachusetts: MIT Press, 2018, p. 29.

117. Project Drawdown, "Public Transport," https://www.drawdown.org/solutions/public-transit (Accessed June 12, 2020).

118. Mike Maciag, "More Americans Now Telecommute Than Take Public Transportation to Work," *Governing*, September 21, 2018, http://www.governing.com/topics/transportation-infrastructure/gov-workers-telework-public-transportation-commute.html.

119. US Census, "MDAT," https://data.census.gov/mdat/#/ (Accessed June 2, 2020).

120. Richard Florida, "The Great Divide in How Americans Commute to Work," CityLab, January 22, 2019, https://www.citylab.com/transportation/2019/01/commuting-to-work-data-car-public-transit-bike/580507/.

121. Adele Peters, "This New Simulator Helps Cities Test a Future of On-demand Transit," *Fast Company*, August 1, 2017, https://www.fastcompany.com/40446331/this-new-simulator-helps-cities-test-a-future-of-on-demand-transit.

122. Jason Laughlin, "As Uber Grows, Septa to Rethink Bus Service," *Philadelphia Inquirer,* July 23, 2017, http://www.philly.com/philly/business/transportation/as-uber-grows-septa-to-rethink-bus-service-20170721.html?arc404=true.

123. Uber, "What Is Uber Pool," https://www.uber.com/ride/uberpool/ (Accessed September 15, 2018).

124. Car2Go, "Pricing in Seattle," https://www.car2go.com/US/en/seattle/how/ (Accessed June 5, 2020).

125. Seattle Department of Transportation, "Seattle Car Share Program Update," March 4, 2016, http://sdotblog.seattle.gov/2016/03/04/seattle-car-share-program-update/.

126. Ibid.

127. Alison Moodie, "US Car Sharing Service Kept 28,000 Private Cars Off the

Road in 3 Years," *The Guardian*, July 23, 2016, https://www.theguardian.com/sustainable-business/2016/jul/23/car-sharing-helps-environment-pollution.

128. Zipcar, "Zipcar: Driving Change in 2018," https://www.zipcar.com/impact (Accessed June 5, 2020).

129. Adam Cohen, video interview with author, June 12, 2020.

130. Emily Badger, "Is Uber Helping or Hurting Mass Transit," *New York Times*, October 16, 2017, https://www.nytimes.com/2017/10/16/upshot/is-uber-helping-or-hurting-mass-transit.html.

131. Robert Hampshire et al., "Measuring the Impact of an Unanticipated Disruption of Uber/Lyft in Austin, TX," May 31, 2017, https://ssrn.com/abstract=2977969 or http://dx.doi.org/10.2139/ssrn.2977969.

132. Zipcar, 2018.

133. Zipcar, 2018.

134. Scott Page, *The Difference: How the Power of Diversity Creates Better Groups, Firms, Schools and Societies*, Princeton, New Jersey: Princeton University Press, 2008, p. 77.

135. Daniel Kahneman, *Thinking, Fast and Slow*, New York: Farrar, Straus and Giroux, 2011, p. 217.

136. Kahneman, p. 219.

137. Matt Ridley, *The Rational Optimist: How Prosperity Evolves*, New York: HaperCollins Publishers, 2010, p. 107.

138. Uber, "Nairobi, Your UberCHAPCHAP Is Arriving Now," Uber Blog, January 17, 2018, https://www.uber.com/en-KE/blog/nairobi/nairobi-your-uberchapchap-is-arriving-now/.

139. Nat Levy, "Car2Go and ReachNow Car-Sharing Services to Merge in Deal between Auto Giants Daimler, BMW," GeekWire, March 28, 2018, https://www.geekwire.com/2018/car2go-reachnow-car-sharing-services-merge-deal-auto-giants-daimler-bmw/.

140. SHARE NOW, "Important Update: Service Ending February 29th," https://www.share-now.com/us/en/important-update/ (Accessed March 26, 2020).

141. Federal Highway Administration, "Analysis of Travel Choices and Scenarios for Sharing Rides," US Department of Transportation, Publication No. FHWA-HOP-21-011, March 2021, https://ops.fhwa.dot.gov/publications/fhwahop21011/fhwahop21011.pdf.

142. Nick Aster, "Hytch Makes Carpooling Pay in Tennessee," *Triple Pundit*, June 18, 2018, https://www.triplepundit.com/2018/06/hytch-makes-carpooling-pay-tennessee/.

143. Todd Wynn, "Oregon's Carbon Offset Scam," Cascade Policy Institute,

February 3, 2009, http://www.cascadepolicy.org/climate-change/oregons-carbon-offset-scam/.

144. Adele Peters, "This New Simulator Helps Cities Test a Future of On-demand Transit," *Fast Company*, August 1, 2017, https://www.fastcompany.com/40446331/this-new-simulator-helps-cities-test-a-future-of-on-demand-transit.

145. Kenneth Gillingham, "Carbon Calculus: For Deep Greenhouse Gas Emission Reductions, a Long-Term Perspective on Costs Is Essential," *Finance & Development* 56, no. 4 (December 2019), https://www.imf.org/external/pubs/ft/fandd/2019/12/the-true-cost-of-reducing-greenhouse-gas-emissions-gillingham.htm.

146. Greenlines, "Home | Greenlines Technology," https://www.greenlines.cc/ (Accessed July 10, 2020).

147. US National Renewable Energy Laboratory, "Google Taps NREL Expertise to Incorporate Energy Optimization into Google Maps Route Guidance," April 12, 2021, https://www.nrel.gov/news/program/2021/google-taps-nrel-expertise-to-incorporate-energy-optimization-into-google-maps-route-guidance.html.

148. Jeff Gonder (Group Manager: Mobility, Behavior and Advanced Powertrains, National Renewable Energy Laboratory), phone interview with author, June 11, 2021.

149. Google, "Redefining What a Map Can Be with New Information and AI," March 30, 2021, https://blog.google/products/maps/redefining-what-map-can-be-new-information-and-ai/.

150. Jacob Holden, Nicholas Reinicke, and Jeff Cappellucci, "RouteE: A Vehicle Energy Consumption Prediction Engine," *SAE International Journal of Advanced and Current Practices in Mobility*, no. 5 (2020): 2760–7, https://doi.org/10.4271/2020-01-0939.

151. Sarah Kessler, "Cities Say Uber and Lyft Are 'Gap Fillers' in Public Transit," *Fast Company*, March 16, 2016, https://www.fastcompany.com/3057907/cities-say-uber-and-lyft-are-gap-fillers-in-public-transit.

152. Ben Schiller, "33% of US Cities Have a 'Very Poor' Relationship with Sharing Companies," *Fast Company*, November 3, 2017, https://www.fastcompany.com/40491028/33-of-u-s-cities-have-a-very-poor-relationship-with-sharing-companies.

153. Metro, "MicroTransit Pilot Project," https://www.metro.net/projects/microtransit/ (Accessed September 15, 2018).

154. Alan Ohnsma, "Think Small: Uber, Lyft, Ford's Chariot Get Chance at Big Role in LA 'MicroTransit,'" *Forbes*, October 23, 2017, https://www.

forbes.com/sites/alanohnsman/2017/10/23/think-small-uber-lyft-fords-
chariot-get-chance-at-big-role-in-la-microtransit/#52af440e6854.

155. Pantonium, "Pantonium's On-Demand Transit in Belleville Ontario," May 1,
2020, https://www.youtube.com/watch?v=bTEHvvTgvVs.

156. Galen Simmons, "Stratford's Weekend On-demand Transit Program Reducing Wait and Ride Times for Users and Fuel Costs for the City," *Stratford
Beacon Herald*, March 26, 2021, https://www.stratfordbeaconherald.com/
news/local-news/stratfords-weekend-on-demand-transit-program-reducing-
wait-and-ride-times-for-users-and-fuel-costs-for-the-city.

157. Pantonium (@Pantoniuminc), "With on-demand Macrotransit, the city has
improved passenger convenience while reducing operating costs. Results after
six weeks: • Buses operating: From 9 [down] to 6 • 81% of trips 30 minutes or
less • 93% of trips 5 mins or less of quoted time • No transfers for all trips,"
Twitter, April 19, 2021, https://twitter.com/pantoniuminc/status/13842133
12469049345?s=20.

158. Pantonium, "Pantonium Awarded $2 million from Sustainable Development
Technology Canada to Develop On-Demand Transit Software," June 18, 2020,
https://pantonium.com/pantonium-awarded-2-million-from-sustainable-
development-technology-canada-to-develop-on-demand-transit-software/.

159. Michelle Baruchman, "'Major Area' of Seattle Could Forbid Most Cars
under City's New, Greener Transportation Plan," *Seattle Times*, March 18,
2021, https://www.seattletimes.com/seattle-news/transportation/seattle-
releases-goals-to-electrify-transportation-system/?utm_medium.

160. Kim Willsher, "Macron Scraps Fuel Tax Rise in Face of Gilets Jaunes
Protests," *The Guardian*, December 5, 2018, https://www.theguardian.com/
world/2018/dec/05/france-wealth-tax-changes-gilets-jaunes-protests-
president-macron.

161. Kristine Ehrich and Julie Irwin, "Willful Ignorance in the Request
for Product Attribute Information," *Journal of Marketing Research* 42
(August 2005): 266–77, http://journals.ama.org/doi/abs/10.1509/
jmkr.2005.42.3.266?code=amma-site.

CHAPTER 6

162. US Environmental Protection Agency, "Understanding Global Warming
Potentials," https://www.epa.gov/ghgemissions/understanding-global-
warming-potentials (Accessed November 5, 2021).

163. Winston Choi-Schagrin, "Wildfires Are Ravaging Forests Set Aside to Soak
Up Greenhouse Gases," *New York Times*, August 23, 2021, https://www.

nytimes.com/2021/08/23/us/wildfires-carbon-offsets.html#:~:text=
California%E2%80%99s%20carbon%20offset%20program%20works%20
by%20paying%20landowners,additional%20carbon%20being%20stored%20
in%20forests%20like%20these.

164. Terrapass, "Projects We Support," https://terrapass.com/projects/project-list
    (Accessed November 5, 2021).

165. Adele Peters, "In China, You Can Track Your Chicken on—You Guessed
    It—the Blockchain," *Fast Company*, January 12, 2018, https://www.fastcom-
    pany.com/40515999/in-china-you-can-track-your-chicken-on-you-guessed-
    it-the-blockchain.

166. Matthew Hutson, "Can Bitcoin's Cryptographic Technology Help Save
    the Environment?" *Science*, May 22, 2017, http://www.sciencemag.org/
    news/2017/05/can-bitcoin-s-cryptographic-technology-help-save-
    environment.

167. Laura He, "Insurtech Giant Zhongan Plans to Use Facial Recognition,
    Blockchain to Monitor Chickens," *South China Morning Post*, December 10,
    2017, https://www.scmp.com/business/companies/article/2123567/
    blockchain-and-facial-recognition-zhongan-techs-recipe-changing?mc_
    cid=e16c11e2dc&mc_eid=4774ed2c3e.

168. Provenance, "Blockchain: The Solution for Transparency in Product Supply
    Chains," November 21, 2015, https://www.provenance.org/whitepaper.

169. Provenance, "From Shore to Plate: Tracking Tuna on the Blockchain," July
    15, 2016, https://www.provenance.org/tracking-tuna-on-the-blockchain#
    blockchains.

170. Grass Roots Farmers' Cooperative, "How We're Using Blockchain Tech
    for Total Transparency," August 1, 2017, https://grassrootscoop.com/
    blockchain-tech-for-total-transparency/.

171. Tania Snugg, "Galapagos Islands: 'Protection Strategy' Set Up after
    'Hundreds of Chinese Fishing Vessels' Spotted Nearby," *Yahoo! News*, July 28,
    2020, https://uk.news.yahoo.com/galapagos-islands-protection-strategy-set-
    hundreds-chinese-fishing-104500907.html.

172. United Nations Development Programme, "How Blockchain Has Trans-
    formed the Lives of Ecuadorean Cocoa Farmers," January 21, 2020, https://
    medium.com/@UNDP/how-blockchain-has-transformed-the-lives-of-
    ecuadorean-cocoa-farmers-1c89941f549c.

173. FairChain, "Dashboard Home-carousel 1 Personal," https://issuer.fairchain.
    org/fairchain/feedbackTransfered/theotherbarpound/dG15ZXJjzQHdhc2
    hpbmd0b25wb2xpY3kub3Jn.

174. UN Climate Change Conference UK 2021, "Glasgow Leaders' Declaration on Forests and Land Use," November 11, 2021, https://ukcop26.org/ glasgow-leaders-declaration-on-forests-and-land-use/.

175. Guido Van Staveren, video interview with author, June 18, 2021.

176. Todd Myers and Steven Sexton, "A Contrarian View of the Local Food Movement," *Wall Street Journal*, July 16, 2015, https://blogs.wsj.com/ experts/2015/07/16/a-contrarian-view-of-the-local-food-movement/.

177. Michael Nielsen, *Reinventing Discovery*, Princeton, New Jersey: Princeton University Press, 2013, p. 64.

CHAPTER 7

178. Gretchen Bakke, *The Grid: The Fraying Wires Between Americans and Our Energy Future*, New York: Bloomsbury, 2016, p. 196.

179. Scott Kessler, phone interview with author, July 10, 2019.

180. Diane Cardwell, "Solar Experiment Lets Neighbors Trade Energy among Themselves," *New York Times*, March 13, 2017, https://www.nytimes. com/2017/03/13/business/energy-environment/brooklyn-solar-grid-energy-trading.html?_r=0.

181. Ben Zycher, "Subsidizing the Rich through California's Solar Scheme," *Forbes*, January 15, 2016, https://www.forbes.com/sites/realspin/2016/01/15/ california-solar-subsidy-net-metering/#96f4bd4722f9.

182. Michael Lee, video interview with author, March 15, 2021.

183. Yates Electrical Services, press release, July 15, 2017, https://www.facebook. com/YatesElectrical/posts/1885881755006949.

184. Katia Moskvitch, "Estonia May Actually Have a Use for the Blockchain: Green Energy," *Wired*, October 25, 2018, https://www.wired.co.uk/article/ blockchain-energy-renewables-estonia-tokenisation.

185. United Nations Department of Economic and Social Affairs, "UN Awards US $1 million for Bangladesh Solar Entrepreneurship Initiative to Aid Rural Electrification," November 21, 2017, https://www.un.org/development/ desa/en/news/sustainable/2017-energy-grant-winners.html.

186. SOLshare, "About Us—ME SOLshare," https://www.me-solshare.com/ (Accessed July 5, 2019).

187. Nobel Foundation, "The Sveriges Riksbank Prize in Economic Sciences in Memory of Alfred Nobel 1991," October 15, 1991, https://www.nobelprize. org/prizes/economic-sciences/1991/press-release/.

188. R. H. Coase, "The Nature of the Firm," *Economica* 4, no. 16

(November 1937): 386–405, https://onlinelibrary.wiley.com/doi/full/10.1111/j.1468-0335.1937.tb00002.x.

189. R. H. Coase, "The Problem of Social Cost," *Journal of Law and Economics* 3 (October 1960): 1–44, https://www.law.uchicago.edu/files/file/coase-problem.pdf.

190. Cassie Martin, "'The Poisoned City' Chronicles Flint's Water Crisis," *Science News*, July 17, 2018, https://www.sciencenews.org/article/poisoned-city-chronicles-flint-water-crisis.

191. US Environmental Protection Agency Office of Inspector General, "Drinking Water Contamination in Flint, Michigan, Demonstrates a Need to Clarify EPA Authority to Issue Emergency Orders to Protect the Public," October 20, 2016, https://www.epa.gov/sites/production/files/2016-10/documents/_epaoig_20161020-17-p-0004.pdf.

192. US House Committee on Oversight and Government Reform, "Examining Federal Administration of the Safe Drinking Water Act in Flint, Michigan, Part 2: Documents for the Record," March 15, 2016, https://oversight.house.gov/wp-content/uploads/2016/03/Documents-for-the-Record.pdf.

193. Anna Clark, *The Poisoned City: Flint's Water and the American Urban Tragedy*, New York: Metropolitan Books, 2018, p. 120.

194. Anna Clark, 2018, p. 147.

195. Devin Henry, "Former EPA Official: Agency Did Nothing Wrong in Flint," *The Hill*, March 15, 2016, http://thehill.com/policy/energy-environment/273058-former-epa-official-agency-did-nothing-wrong-in-flint.

196. Calestous Juma, *Innovation and Its Enemies*, New York: Oxford University Press, 2016, p. 23.

197. Morgan Clendaniel, "Looking Back at the World Changing Ideas of 2015," *Fast Company*, December 21, 2015, https://www.fastcompany.com/3054216/looking-back-at-the-world-changing-ideas-of-2015.

198. Elinor Ostrom, *Governing the Commons: The Evolution of Institutions for Collective Action*, New York: Cambridge University Press, 1990, p. 214.

199. Michael J. Coren, "Utility Startups Are Making the Electric Grid Work More Like the Internet," Quartz, June 1, 2017, https://qz.com/995551/the-startup-drift-is-one-of-new-yorks-newest-utilities-allowing-peer-to-peer-trading-for-the-electric-smart-grid/.

200. US Energy Information Administration, "Direct Federal Financial Interventions and Subsidies in Energy in Fiscal Year 2016," April 2018, https://www.eia.gov/analysis/requests/subsidy/pdf/subsidy.pdf.

201. Roland Irle, "USA Plug-in Sales for the First Half of 2019," http://www. ev-volumes.com/country/usa/ (Accessed August 2, 2019).

202. Tesla, "On the Road," https://www.tesla.com/supercharger (Accessed August 2, 2019).

203. Suzanne Guinn, "EVSE Rebates and Tax Credits, by State," November 2, 2018, https://www.clippercreek.com/evse-rebates-and-tax-credits-by-state/.

204. Kristen Hall-Geisler, "ChargePoint Creates Waitlist for Public EV Charging," September 12, 2016, https://techcrunch.com/2016/09/12/chargepoint-creates-waitlist-for-public-ev-charging/.

205. International Union for the Conservation of Science, "*Dermochelys coriacea* (leatherback)," https://www.iucnredlist.org/species/6494/43526147 (Accessed December 7, 2019).

206. US Fish and Wildlife Service, "Leatherback Sea Turtle," https://www.fws. gov/northflorida/SeaTurtles/Turtle%20Factsheets/leatherback-sea-turtle.htm (Accessed June 10, 2018).

207. Felipe D'Amoed, phone interview with author, January 19, 2018.

208. Wendy Purnell, email interview with author, January 12, 2018.

209. Meagan Treacy, "Decoy Sea Turtle Eggs Help Track Down Poachers," Tree-hugger, October 4, 2017, https://www.treehugger.com/clean-technology/decoy-sea-turtle-eggs-help-track-down-poachers.html.

210. Sarah Birnbaum, "What I Learned from 'Breaking Bad' about Saving Sea Turtles," PRI, June 5, 2017, https://www.pri.org/stories/2017-06-05/what-i-learned-breaking-bad-about-saving-sea-turtles.

211. Jeremy Hance, "How to Catch a Poacher: Breaking Bad and Fake Eggs," *The Guardian*, July 26, 2016, https://www.theguardian.com/environment/radical-conservation/2016/jul/26/sea-turtles-breaking-bad-hollywood-nicaragua-poaching-wildlife-trade.

212. InvestEGGator, "USAID's Wildlife Crime Tech Challenge," https:// investeggator.com/wildlife-crime-tech-challenge/ (Accessed June 19, 2018).

213. Jacob Byl, "Perverse Incentives and Safe Harbors in the Endangered Species Act: Evidence from Timber Harvests Near Woodpeckers," *Ecological Economics* 157 (March 2019): 100–108, https://www.sciencedirect.com/science/article/pii/S0921800918309789#! (Accessed August 3, 2019).

214. Johnathan Adler, "Perverse Incentives and the Endangered Species Act," *Resources*, August 4, 2008, https://www.resourcesmag.org/common-resources/perverse-incentives-and-the-endangered-species-act/.

215. US Fish and Wildlife Service, "ESA Basics: 40 Years of Conserving Endangered Species," February 2017, https://www.fws.gov/endangered/

esa-library/pdf/ESA_basics.pdf.

216. Washington State Department of Ecology, "Wetland Mitigation Banking," https://ecology.wa.gov/Water-Shorelines/Wetlands/Mitigation/Wetland-mitigation-banking (Accessed August 3, 2019).

CHAPTER 8

217. Long Live the Kings, "Fish Don't Leave Footprints: Steelhead Tracking 101," March 22, 2017, https://lltk.org/fish-dont-have-footprints-steelhead-tracking-101/.

218. Although steelhead are a type of trout, because they spend much of their life in saltwater, they are frequently referred to as salmonids.

219. Dahlia Bazzaz, "Salmon Migration Lessons Like Fantasy Football for Fish, with 2,000 Classrooms Playing," *Seattle Times*, May 17, 2018, https://www.seattletimes.com/education-lab/salmon-migration-lessons-like-fantasy-football-for-fish-with-2000-classrooms-playing/.

220. Aldo Leopold, "Wildlife in American Culture," *Journal of Wildlife Management* 7, no. 1 (1943): 1–6, www.jstor.org/stable/3795771.

221. Roger Scruton, *How to Think Seriously about the Planet: The Case for an Environmental Conservatism*, New York: Oxford University Press, 2012, p. 252.

222. Catherine Early, "The Spy Tech That Listens Out for Illegal Logging," Five Media, April 6, 2021, https://fivemedia.com/articles/the-spy-tech-that-listens-out-for-illegal-logging/.

223. Christian Nellemann et al., eds., *The Environmental Crime Crisis—Threats to Sustainable Development from Illegal Exploitation and Trade in Wildlife and Forest Resources*, 2014, p. 18, https://reliefweb.int/sites/reliefweb.int/files/resources/RRAcrimecrisis.pdf.

224. Christina Nunez, "Your Old Cell Phone Can Help Save the Rain Forest," *National Geographic*, June 15, 2017, https://www.nationalgeographic.com/adventure/article/topher-white-engineer-rainforests-explorer-festival.

225. Chrissy Durkin, video interview with author, June 18, 2021.

226. Kickstarter, "Rainforest Connection—Phones Turned to Forest Guardians by Rainforest Connection," Kickstarter, https://www.kickstarter.com/projects/topherwhite/rainforest-connection-phones-turned-to-forest-guar/description.

227. Jackie Snow, "Better Data, Cheaper Tech Promise to Unlock Nature's Secrets," *Wall Street Journal*, February 8, 2021, https://www.wsj.com/articles/better-data-cheaper-tech-promise-to-unlock-natures-secrets-11612800044.

228. Tony Iwa, phone interview with author, May 8, 2018.

229. Grant Van Horn, video interview with author, June 16, 2021.

230. Grant Van Horn et al., "The INaturalist Species Classification and Detection Dataset," Proceedings of the IEEE Conference on Computer Vision and Pattern Recognition (CVPR), 2018, pp. 8769–78.

231. Scott Loarie, video interview with author, July 7, 2021.

232. Geoffrey Cox, "What Is a 'Verifiable Observation' and How Does It Reach 'Research Grade'?" iNaturalist, August 6, 2019, https://www.inaturalist.org/posts/26549-what-is-a-verifiable-observation-and-how-does-it-reach-research-grade.

233. Global Biodiversity Information Facility, "What is GBIF?" https://www.gbif.org/what-is-gbif.

234. California Academy of Sciences, "Global Nature Observation Network iNaturalist Surpasses 50 Million Wild Plant and Animal Observations," September 21, 2020, https://www.calacademy.org/press/releases/inaturalist-50-million.

235. iNaturalist's blog, "One Sixth of All Named Species Tallied!" June 2, 2021, https://www.inaturalist.org/blog/52872-one-sixth-of-all-named-species-tallied.

236. National Academies of Sciences, Engineering, and Medicine; Division of Behavioral and Social Sciences and Education; Board on Science Education; Committee on Designing Citizen Science to Support Science Learning; K. A. Dibner and R. Pandya, eds. "Learning through Citizen Science: Enhancing Opportunities by Design," Washington (DC): National Academies Press (US); November 1, 2018, summary, https://www.ncbi.nlm.nih.gov/books/NBK535953/.

237. SciStarter, "About Us—SciStarter," https://scistarter.org/about (Accessed June 13, 2021).

238. Darlene Cavalier, video interview with author, June 8, 2021.

239. Kathryn Youngblood, video interview with author, June 9, 2021.

240. Mississippi River Plastic Pollution Initiative, "Marine Debris Tracker Citizen Science Field Guide," https://debristracker.org/static/media/Mississippi_River_Citizen_Science_Field_Guide_v6.3ecdc835.pdf (Accessed June 13, 2021).

241. Pattrn (@pattrn), "Scientists with @universityofga are using GPS to follow plastic bottles throughout their journey down the Mississippi River. We got an update from the project's leader. The results are surprising. #plasticpollution #plasticwaste," Twitter, May 28, 2021, https://t.co/4i9BareDHM.

242. Gabriel I. Gadsden et al., "Michigan ZoomIN: Validating Crowd-Sourcing to Identify Mammals from Camera Surveys," *Wildlife Society Bulletin*, no. 2 (June 2021): 221–9, https://doi.org/10.1002/wsb.1175.

243. Microsoft Camera Traps, "CameraTraps/megadetector.md at master • microsoft/CameraTraps • GitHub," https://github.com/microsoft/ CameraTraps/blob/master/megadetector.md (Accessed June 11, 2021).

244. iWildCam, "GitHub—visipedia/iwildcam_comp: iWildCam competition details," https://github.com/visipedia/iwildcam_comp (Accessed June 11, 2021).

245. iWildCam, "iwildcam_comp/2020 at master · visipedia/iwildcam_comp · GitHub," https://github.com/visipedia/iwildcam_comp/tree/master/2020 (Accessed June 11, 2021).

246. Caren B. Cooper, Jennifer Shirk, and Benjamin Zuckerberg, "The Invisible Prevalence of Citizen Science in Global Research: Migratory Birds and Climate Change," PLoS ONE 9, no. 9 (2014): e106508, https://doi.org/ 10.1371/journal.pone.0106508.

247. Leslie Ries and Karen Oberhauser, "A Citizen Army for Science: Quantifying the Contributions of Citizen Scientists to Our Understanding of Monarch Butterfly Biology," *BioScience* 65, no. 4 (2015): 419–30, JSTOR, www.jstor. org/stable/90007262.

248. George MacKerron and Susana Mourato, "Happiness is Greater in Natural Environments," *Global Environmental Change* 23, no. 5 (2013): 992–1000, http://eprints.lse.ac.uk/49376/.

249. Joel Methorst et al., "The Importance of Species Diversity for Human Well-being in Europe," *Ecological Economics* 181 (2021): 106917, https://doi. org/10.1016/j.ecolecon.2020.106917.

250. Florence Willis, *The Nature Fix*, New York: W. W. Norton & Company, 2017, p. 100.

251. Shawn Regan, "Want Public Access to Private Land? There's an App for That," *High Country News*, January 28, 2016.

252. Hipcamp.com, "About Hipcamp | Getting More People Outside with Unique Outdoor Stays," https://www.hipcamp.com/en-US/about#our-story.

253. onX, "Morel Mushroom Maps with the onX Hunt App," April 1, 2020, https://www.onxmaps.com/hunt/blog/morel-mushroom-hunting-maps- where-to-find.

254. Russ McSpadden (@PeccaryNotPig), "A coyote and a badger use a culvert as a wild-life crossing to pass under a busy California highway together. Coyotes and badgers are known to hunt together. Peninsula Open Space Trust," Twitter, February 3, 2020, https://twitter.com/PeccaryNotPig/status/1224515892282740737?s=20.

CHAPTER 9

255. Cyber Squirrel 1, "CyberSquirrel1.com," https://cybersquirrel1.com/.

256. Jason Breslow, "20 Companies Are Behind Half of the World's Single-Use

Plastic Waste, Study Finds," NPR, May 18, 2021, https://www.npr.org/2021/05/18/997937090/half-of-the-worlds-single-use-plastic-waste-is-from-just-20-companies-says-a-stu.

257. Jaime Nack, "5 Ways You Can Personally Fight the Climate Crisis," World Economic Forum, May 27, 2019, https://www.weforum.org/agenda/2019/05/5-ways-you-can-personally-fight-the-climate-crisis-change-global-warming/.

258. James Wright, "On BBC Two, George Monbiot Had the Entire Studio Applauding the Overthrow of Capitalism," The Canary, April 12, 2019, https://www.thecanary.co/trending/2019/04/12/on-bbc-two-george-monbiot-had-the-entire-studio-applauding-the-overthrow-of-capitalism/.

259. Damien Cave, "It Was Supposed to Be Australia's Climate Change Election. What Happened?" New York Times, May 19, 2019, https://www.nytimes.com/2019/05/19/world/australia/election-climate-change.html.

260. Josh Wingrove, "Doug Ford Wins Ontario Vote on Populist Message," Detroit News, June 8, 2018, https://www.detroitnews.com/story/news/world/2018/06/08/doug-ford-wins-ontario/35836615/.

261. Jamey Keaten, "Swiss Narrowly Reject Tax Hike to Fight Climate Change," Associated Press, June 12, 2021, https://news.yahoo.com/swiss-vote-whether-hike-taxes-065833595.html.

262. Climate Action Tracker, "Countries | Climate Action Tracker," https://climateactiontracker.org/countries/ (Accessed June 15, 2021).

263. Charles Darwin, The Origin of Species, New York: Barnes and Noble Classics, 2004, pp. 69–70.

264. BirdNET, "BirdNET LiveMap (cornell.edu)," https://birdnet.cornell.edu/map (Accessed June 15, 2021).

265. Curtis Tongue, video interview with author, July 1, 2021.

266. Matt Ridley, "Don't Write Off the Next Big Thing Too Soon," The Times, November 6, 2017, https://www.thetimes.co.uk/article/dont-write-off-the-next-big-thing-too-soon-rbf2q9sck.

267. Arne Holst, "Number of Internet of Things (IoT) Connected Devices Worldwide from 2019 to 2030," Statista, January 20, 2021, https://www.statista.com/statistics/1183457/iot-connected-devices-worldwide/.

268. Pacific Northwest National Laboratory, "Internet of Things Common Operating Environment," https://www.pnnl.gov/internet-things-common-operating-environment (Accessed June 23, 2021).

269. Christine Fisher, "Philips Patched a Longstanding Hue Bulb Security Flaw," Endgadget, February 5, 2020, https://www.engadget.com/2020-02-05-philips-hue-signify-vulnerability.html.

270. Charles DeBeck, "I Can't Believe Mirais: Tracking the Infamous IoT Malware," Security Intelligence, July 18, 2019, https://securityintelligence.com/posts/i-cant-believe-mirais-tracking-the-infamous-iot-malware-2/.

271. Rich Castagna, "Energy Grid Security Gets More Challenging with IoT," IoT World Today, August 18, 2020, https://www.iotworldtoday.com/2020/08/18/energy-grid-security-gets-more-challenging-with-iot/.

272. Geoff Earle, "Biden Threatens Cyber War on Russia: US President Promises 'Retaliation' if Putin Attacks a List of 16 'Critical' Facilities—Then Snaps at Press and Says He Is NOT Confident the Russian President Will Change," Daily Mail, June 16, 2021, https://www.dailymail.co.uk/news/article-9694963/Biden-threatens-cyber-response-against-Putin-gives-list-16-no-hack-infrastructure-locations.html.

273. Stu Steiner, video interview with author, June 22, 2021.

274. Cyberforce Competition, "CyberForce Competition—Department of Energy's CyberForce® Program," https://cyberforcecompetition.com/cyberforce/ (Accessed June 24, 2021).

275. PISCES, "About—PISCES," https://pisces-intl.org/about/.

276. Pacific Northwest National Labs, "Internet of Things Common Operating Environment | PNNL," https://www.pnnl.gov/internet-things-common-operating-environment.

277. Pacific Northwest National Laboratory, "Internet of Things Common Operating Environment," PNNL-SA-151557, April 2020, https://www.pnnl.gov/sites/default/files/media/file/Internet%20of%20Things%20Common%20Operating%20Environment.pdf.

278. Eunsun Cho, "The Social Credit System: Not Just Another Chinese Idiosyncrasy," Journal of Public & International Affairs, May 1, 2020, https://jpia.princeton.edu/news/social-credit-system-not-just-another-chinese-idiosyncrasy.

279. Matt Dougherty, "Woke Up Sweating: Some Texans Shocked to Find Their Smart Thermostats Were Raised Remotely," KHOU-TV, June 17, 2021, https://www.khou.com/article/news/local/texas/remote-thermostat-adjustment-texas-energy-shortage/285-5acf2bc5-54b7-4160-bffe-1f9a5ef4362a.

280. Zack Whittaker, "Ring Refuses to Say How Many Users Had Video Footage Obtained by Police," TechCrunch, June 8, 2021, https://techcrunch.com/2021/06/08/ring-police-warrants-neighbors/.

281. Ring, "Ring Launches Request for Assistance Posts on the Neighbors App," June 3, 2021, https://blog.ring.com/products-innovation/ring-launches-request-for-assistance-posts-on-the-neighbors-app/ (Accessed June 27, 2021).

282. Chris Sandbrook et al., "Principles for the Socially Responsible Use of

Conservation Monitoring Technology and Data," *Conservation Science and Practice* 3 (2021): e374. https://doi.org/10.1111/csp2.374.

283. WILDLABS, "How do I use conservation tech ethically? | WILDLABS. NET," Wildlabs.net, July 26, 2021, https://www.wildlabs.net/resources/tech-tutors/how-do-i-use-conservation-tech-ethically.

284. eBird, "Sensitive Species in eBird," November 16, 2017, https://ebird.org/news/sensitive-species-in-ebird.

285. eBird, "Sensitive Species in eBird," April 30, 2021, https://support.ebird.org/en/support/solutions/articles/48000803210-sensitive-species-in-ebird (Accessed June 27, 2021).

286. WWF, "WWF Wildlife Cyber Spotter Program," https://www.worldwildlife.org/pages/wwf-wildlife-cyber-spotter-program (Accessed June 27, 2021).

CHAPTER 10

287. Jay Inslee and Bracken Hendricks, *Apollo's Fire*, Washington: Island Press, 2008, p. 303.

288. US Environmental Protection Agency, "Overview for Renewable Fuel Standard | US EPA," https://www.epa.gov/renewable-fuel-standard-program/overview-renewable-fuel-standard (Accessed August 26, 2021).

289. US Environmental Protection Agency, "Annual Compliance Data for Obligated Parties and Renewable Fuel Exporters under the Renewable Fuel Standard (RFS) Program | US EPA," https://www.epa.gov/fuels-registration-reporting-and-compliance-help/annual-compliance-data-obligated-parties-and (Accessed August 26, 2021).

290. Jenna R. Jambeck et al., "Plastic Waste Inputs from Land into the Ocean," *Science* 347, no. 6223 (February 13, 2015): 768–71, https://doi.org/10.1126/science.1260352.

291. World Economic Forum, "More Plastic than Fish in the Ocean by 2050: Report Offers Blueprint for Change," January 19, 2016, https://www.weforum.org/press/2016/01/more-plastic-than-fish-in-the-ocean-by-2050-report-offers-blueprint-for-change.

292. The Ocean Cleanup, "The Ocean Cleanup," https://theoceancleanup.com/ (Accessed July 5, 2021).

293. #TeamSeas, "#TeamSeas," https://teamseas.org/.

294. The Ocean Cleanup, "The Numbers Behind Our Catch," May 20, 2021, https://theoceancleanup.com/updates/the-numbers-behind-our-catch/.

295. Erik Stokstad, "Controversial Plastic Trash Collector Begins Maiden Ocean Voyage," *Science*, September 11, 2018, https://www.sciencemag.org/news/

2018/09/still-controversial-plastic-trash-collector-ocean-begins-maiden-voyage.

296. Shaun Frankson, phone interview with author, May 20, 2019.

297. Plastic Bank, "Frequently Asked Questions," https://plasticbank.com/faq/ (Accessed July 6, 2021).

298. Plastic Bank, "Plastic Bank Is #1 for Certified Environment & Social Impact," https://plasticbank.com/our-impact/ (Accessed July 6, 2021).

299. Insureblocks, "Ep. 158—Deep Dive on Plastic Bank's Blockchain," April 25, 2021, https://insureblocks.com/ep-158-deep-dive-on-plastic-banks-blockchain/.

300. UK Office of National Statistics, "Atmospheric Emissions: Greenhouse Gases by Industry and Gas," https://www.ons.gov.uk/economy/environmentalaccounts/datasets/ukenvironmentalaccountsatmosphericemissionsgreenhousegas emissionsbyeconomicsectorandgasunitedkingdom (Accessed July 11, 2021).

301. National Grid ESO, "Zero-carbon Explained | National Grid ESO," https://www.nationalgrideso.com/electricity-explained/zero-carbon-explained (Accessed July 15, 2021).

302. Phil Steele, phone interview with author, January 17, 2020.

303. Octopus Energy, "Octopus Energy Group Reports FY19/20 Fiscal Results," May 7, 2021, https://octopus.energy/press/octopus-energy-group-reports-fy1920-fiscal-results/.

304. Maddie Inglis, "With AgileOctopus, the UK's Greenest Energy Was Also the Cheapest | Octopus Energy," Octopus Energy Blog, December 10, 2019, https://octopus.energy/blog/cheaper-greener-agile-energy/.

305. Duncan Burt (@DBBurt), "A big thank you to all those #EV drivers and smart cookies, including everyone on #OctopusAgile, who helped us balance the GB grid last night. Getting paid to use more power on a windy night! @octopus_energy @enappsys @ng_eso," Twitter, December 8, 2019, https://twitter.com/DBBurt/status/1203636331924115456.

306. Octopus Energy, "Agile Octopus: A Consumer-led Shift to a Low Carbon Future," https://octopus.energy/static/consumer/documents/agile-report.pdf.

307. Maddie Inglis, "A Hack Day for the Future of Energy | Octopus Energy," August 13, 2018, https://octopus.energy/blog/hacking-the-future-of-energy/.

308. Mick Wall, email to author, March 11, 2020.

309. Kim Bauters, video interview with author, March 3, 2020.

310. Jan Rosenow, video interview with author, March 4, 2020.

311. UK Office of Gas and Electricity Markets, "Smart Export Guarantee (SEG) | Ofgem," https://www.ofgem.gov.uk/environmental-and-social-schemes/smart-export-guarantee-seg (Accessed July 17, 2021).

312. Solar Panel Prices, "The Best SEG/Export Tariffs in 20–League Table | Solar Panel Prices," June 5, 2021, https://www.solarpanelprices.co.uk/articles/solar-panels/best-smart-export-guarantee-tariffs/ (Accessed July 17, 2021).

313. Octopus Energy, "Introducing Octopus Fan Club: Local Green Energy for Your Home | Octopus Energy," https://octopus.energy/octopus-fan-club/ (Accessed July 17, 2021).

314. Michael Lee, phone interview with author, December 2019.

315. US Energy Information Administration, "Electric Power Monthly with Data for February 2021," US Department of Energy, April 2021, https://www.eia.gov/electricity/monthly/archive/april2021.pdf, p. 138.

CHAPTER 11

316. Michael P. Hall, Neil A. Lewis, and Phoebe C. Ellsworth, "Believing in Climate Change, But Not Behaving Sustainably: Evidence from a One-year Longitudinal Study," *Journal of Environmental Psychology* 56 (2018): 55–62, https://doi.org/10.1016/j.jenvp.2018.03.001.

317. Montana Department of Natural Resources and Conservation, "Montana Statewide Assessment of Forest Conditions, Draft V9.0 9/25/2020," September 9, 2020, p. 7, http://dnrc.mt.gov/public-interest/environmental-docs/2020/october/montana-statewide-assessment-of-forest-conditions.

318. See, for example, Montana Department of Natural Resources and Conservation, "Montana Statewide Assessment of Forest Conditions, Draft V9.0 9/25/2020," September 9, 2020, p. 58, http://dnrc.mt.gov/public-interest/environmental-docs/2020/octobermontana-statewide-assessment-of-forest-conditions.

319. United Nations Educational, Scientific, and Cultural Organization, "Indigenous Peoples: Sustainable Development and Environmental Change," https://en.unesco.org/indigenous-peoples/sustainable-development (Accessed August 1, 2021).

320. Mark Sagoff, "What Would Hayek Do about Climate Change?" Breakthrough Institute, March 1, 2021, https://thebreakthrough.org/journal/no-13-winter-2021/what-would-hayek-do-about-climate-change.

321. Sam Kazman, "Drug Approvals and Deadly Delays," *Journal of American Physicians and Surgeons* 15, no. 4 (Winter 2010), https://www.jpands.org/vol15no4/kazman.pdf.

322. National Snow & Ice Data Center, "Charctic Interactive Sea Ice Graph | Arctic Sea Ice News and Analysis (nsidc.org)," http://nsidc.org/arcticseaicenews/charctic-interactive-sea-ice-graph/.

323. California ISO, "California ISO (caiso.com)," http://www.caiso.com/Pages/default.aspx.

324. Roy Spencer (drroyspencer.com), "Latest Global Temperatures," https://www.drroyspencer.com/latest-global-temperatures/.

325. Goddard Institute for Space Studies, "Data.GISS: GISS Surface Temperature Analysis (v4): Analysis Graphs and Plots (nasa.gov)," https://data.giss.nasa.gov/gistemp/graphs_v4/.

326. Crowdsourcing and Citizen Science Act, 15 US Code 3724, https://uscode.house.gov/view.xhtml?req=granuleid:USC-prelim-title15-section3724&num=0&edition=prelim.

327. EarthRanger, Skylight, and **WILD**LABS, "The Future of Conservation: How Technology Is Transforming Wildlife Conservation during COVID-19," August 21, 2021, https://earthranger.com/custom/pdf/The_Future_of_Conservation_Report.pdf.

328. America250 Foundation, "Meet the Co-chairs: Innovation, Science, and Entrepreneurship Advisory Council—America250," America250.org, August 12, 2021, https://america250.org/story/meet-the-co-chairs-innovation-science-and-entrepreneurship-advisory-council/.

INDEX

ABOUT THE AUTHOR

With more than two decades in environmental policy, Todd Myers's experience includes work on a range of environmental issues, including climate policy, forest health, old-growth forests, and salmon recovery. A former member of the executive team at the Washington State Department of Natural Resources, he is a member of the Puget Sound Salmon Recovery Council and is the environmental director at the Washington Policy Center, a public policy think tank in Seattle.

His writing has appeared in the *Wall Street Journal*, *National Review*, *Seattle Times*, and *USA Today*, and he has appeared on numerous news networks including CNBC, Fox News, the BBC, and CNN. He served as vice president of the Northwest Association of Biomedical Research and received their Distinguished Service Award in 2018 for his support of bioscience. He has also served as president of the Prescription Drug Assistance Foundation, a nonprofit providing medicines to low-income patients.

In 2021, Myers served as president of his local beekeeping club in his quest to build an army of stinging insects at his command. He has a bachelor's degree in politics from Whitman College and a master's degree in Russian/International Studies from the Jackson School of International Studies at the University of Washington. He and his wife, Maria, live in the Cascade Mountains in Washington State with 200,000 honeybees, and he claims to make an amazing pasta carbonara and an incredible dirty vodka martini with blue-cheese-stuffed olives.